Thomas Wolf

Ultrafast photophysics and photochemistry of radical precursors in solution

Ultrafast photophysics and photochemistry of radical precursors in solution

Ultrafast photophysics and photochemistry of radical precursors in solution

by
Thomas Wolf

Dissertation, Karlsruher Institut für Technologie (KIT)
Fakultät für Chemie und Biowissenschaften, 2012
Referenten: PD Dr. A.-N. Unterreiner, Prof. Dr. M. Kappes

Impressum

Karlsruher Institut für Technologie (KIT)
KIT Scientific Publishing
Straße am Forum 2
D-76131 Karlsruhe
www.ksp.kit.edu

KIT – Universität des Landes Baden-Württemberg und
nationales Forschungszentrum in der Helmholtz-Gemeinschaft

KIT Scientific Publishing 2013
Print on Demand

ISBN 978-3-86644-940-4

Ultrafast photophysics and photochemistry of radical precursors in solution

Zur Erlangung des akademischen Grades eines

DOKTORS DER NATURWISSENSCHAFTEN

(Dr. rer. nat.)

Fakultät für Chemie und Biowissenschaften

Karlsruher Institut für Technologie (KIT) - Universitätsbereich

genehmigte

DISSERTATION

von

Thomas J. A. Wolf

aus

Filderstadt

Dekan:	Prof. Dr. M. Bastmeyer
Referent:	PD Dr. A.-N. Unterreiner
Korreferent:	Prof. Dr. M. Kappes
Tag der mündlichen Prüfung:	19.10.2012

Summary

The excited states dynamics leading to photoinduced generation of radicals and their dynamics in solution have been investigated in this work by femtosecond transient absorption spectroscopy assisted by quantum chemical calculations. The investigated systems can be sorted into three categories: Fully chlorinated and brominated cyclopentadienes show the fastest dynamics leading to radicals on a femtosecond time scale, which makes the investigation of subsequent reactions possible on a picosecond time scale. The photoinitiators benzoin, 2,4,6-trimethylbenzoin and mesitil react to form radicals from their triplet states via α-cleavage within several picoseconds. The photoinitiators 7-diethylamino-3-thenoylcoumarin and isopropylthioxanthone exhibit the slowest dynamics, and only the intersystem crossing from the excited singlet state into the triplet manifold can be observed on the investigated time scale of 1.6 nanoseconds. In the following, results from all three categories will be discussed in more detail.

Photoinduced dynamics of halogenated cyclopentadienes

The photoinduced dynamics of fully chlorinated (C_5Cl_6) and brominated (C_5Br_6) cyclopentadienes have been investigated in the solvents cyclohexane, isopropanol, chloroform and trichloroethanol. Comparison with data from time-resolved photoelectron spectroscopy in the gas phase shows that both molecules undergo homolytic bond dissociation within < 100 femtoseconds to form a cyclopentadienyl and a halogen radical. In solution the photolytically generated radicals geminately form charge-transfer complexes within several picoseconds after bond dissociation. Two processes competing with generation of the complexes are: Escape of the halogen radical from the solvent cage, and geminate recombination of the radicals taking place within few picoseconds.

In the case of C_5Br_6 the competition is only moderately solvent-dependent. The complexes are observed to be surprisingly stable and exhibit lifetimes in the nanosecond range. Thus, for the first time, the fate of individual charge-transfer complexes of halogen radicals can be investigated.

The competition between complex formation and side processes is highly solvent-dependent in the case of C_5Cl_6. In cyclohexane, for instance, formation of a charge-transfer complex cannot be found. In the solvent trichloroethanol the complex is observed to be less stable by an order of magnitude than in the other solvents, which can be regarded as a hint for a direct H abstraction reaction of the Cl radical with a molecule from the solvent cage.

Early steps in polymerization initiation by type I photoinitiators

The three photoinitiators benzoin, 2,4,6-trimethylbenzoin, and mesitil have been investigated by transient absorption spectroscopy in methanol solution. The photoinduced dynamics of benzoin and 2,4,6-trimethylbenzoin are comparable. Their first excited singlet state is depopulated with time constants of (0.7 ± 0.2) and (1.2 ± 0.2) picoseconds, respectively. The main depopulation channel is intersystem crossing into the triplet manifold. The high efficiency of intersystem crossing can be connected to the special electronic structure of the two photoinitiators, which permits the existence of a triplet state nearly isoenergetic to the excited singlet state. From the lowest triplet state radicals are generated with time constants of (14.3 ± 0.9) and (15.9 ± 0.4) picoseconds by α-cleavage.

Mesitil, in comparison, shows a substantially faster depopulation of the excited singlet state $((0.2 \pm 0.2)$ picoseconds). Due to an electronic structure, which is different from the other two investigated photoinitiators, the depopulation is mainly attributed to internal conversion within the singlet manifold. Intersystem crossing is only a minor depopulation channel. From the lowest triplet state α-cleavage takes place within (25 ± 2) picoseconds.

An estimation of relative intersystem crossing quantum yields nicely reproduces earlier findings from investigations of relative incorporation probabilities. Thus, it can be shown that the overall photoinitiation efficiency is dominated by intersystem crossing in the photoinitiator taking place within the first picoseconds after photoexcitation.

The excited states dynamics of photodepletable triplet photoinitiators

The photoinduced dynamics of the two triplet photoinitiators 7-diethylamino-3-thenoylcoumarin and isopropylthioxanthone, which additionally show fluorescence with considerable quantum yield, have been investigated in ethanol solution. The lifetime of the first excited singlet state of 7-diethylamino-3-thenoylcoumarin is observed to be (99 ± 1) picoseconds. It is depopulated mainly via fluorescence as well as intersystem crossing. The lifetime of the lowest triplet state is beyond the investigated time window. It can be shown that irradiation of molecules being in their excited singlet state with a wavelength of 532 nm predominantly leads to depletion via stimulated emission. Thus, the photoinitiator is highly suitable for stimulated emission depletion lithography.

Isopropylthioxanthone exhibits a substantially longer lifetime of the first excited singlet state $((2.3 \pm 0.2)$ nanoseconds$)$. In general, this would make it more suitable for stimulated emission depletion lithography. However, it can be shown that irradiation of the molecule predominantly leads to excited state absorption rather than stimulated emission in nearly the entire visible spectral range. Since excited state absorption leads to addition of energy to the system instead of removal, the process is counterproductive for stimulated emission depletion lithography. Therefore, the molecule is less suitable for this special application than 7-diethylamino-3-thenoylcoumarin.

Zusammenfassung

Im Rahmen dieser Arbeit wurde die photoinduzierte Dynamik elektronisch angeregter Zustände, welche zur Bildung von Radikalen führt, und die Dynamik der gebildeten Radikale in Lösung mit Hilfe von zeitaufgelöster Absorptionsspektroskopie auf der Femtosekunden- bis Nanosekundenzeitskala untersucht. Zusätzlich wurden zur Interpretation der experimentellen Ergebnisse quantenchemische Rechnungen durchgeführt. Die untersuchten Systeme lassen sich in drei Kategorien einteilen: Perchlorierte und -bromierte Cyclopentadiene zeigen die schnellste Dynamik auf der Femtosekundenzeitskala. Dadurch wird die Beobachtung von Folgereaktionen der gebildeten Radikale auf der Pikosekundenzeitskala möglich. Die Photoinitiatoren Benzoin, 2,4,6-Trimethylbenzoin und Mesitil bilden Radikale innerhalb einiger Pikosekunden durch α-Spaltung aus dem Triplett-Zustand. Die Photoinitiatoren 7-Diethylamino-3-thenoylcumarin und Isopropylthioxanthon weisen die langsamste Dynamik auf, weshalb auf der untersuchten Zeitskala von 1,6 Nanosekunden nur die Singulett-Triplett-Interkombination (intersystem crossing) beobachtet werden kann. Im folgenden werden die Ergebnisse aus den drei Kategorien genauer diskutiert.

Photoinduzierte Dynamik halogenierter Cyclopentadiene

Die photoinduzierte Dynamik perchlorierter und -bromierter Cyclopentadiene wurde in den Lösungsmitteln Cyclohexan, Isopropanol, Chloroform und Trichlorethanol untersucht. Der Vergleich mit Ergebnissen aus Untersuchungen mit Hilfe zeitaufgelöster Photoelektronenspektroskopie zeigt, daß beide Moleküle innerhalb von < 100 Femtosekunden homolytisch in ein Cyclopentadienyl- und ein Halogenradikal dissoziieren. In Lösung bilden die beiden geminal entstandenen Radikale innerhalb einiger Pikosekunden nach Bindungsspaltung Ladungs-Transfer-Komplexe miteinander. Zwei weitere Prozesse, Ausbruch des Halogenradikals aus dem Lösungsmittelkäfig und geminale Rekombination der Radikale, welche beide innerhalb weniger Pikosekunden stattfinden, konkurrieren mit der Bildung des Ladungs-Transfer-Komplexes.

Im Falle von Perbromcyclopentadien (C_5Br_6) ist die Konkurrenz der Prozesse nur leicht lösungsmittelabhängig. Die Komplexe sind mit Lebensdauern im Nanosekundenbereich überraschend stabil. Dadurch kann zum ersten Mal das Schicksal individueller Ladungs-Transfer-Komplexe von Halogenradikalen verfolgt werden.

Im Falle von Perchlorcyclopentadien (C_5Cl_6) ist die Konkurrenz zwischen Komplexbildung und Nebenreaktionen dagegen stark lösungsmittelabhängig. Zum Beispiel kann die Bildung des Ladungs-Transfer-Komplexes in Cyclohexan nicht beobachtet werden. Im Lösungsmittel Trichlorethanol ist eine um eine Größenordnung geringere Stabilität der Ladungs-Transfer-Komplexe von Chlorradikalen als in den anderen Lösungsmitteln beobachtbar. Dies ist ein Hinweis auf eine direkte H-Abstraktionsreaktion des Chlorradikals mit einem Molekül des Lösungsmittelkäfigs.

Frühe Prozesse in der Polymerisations-Startreaktion von Typ-I-Photoinitiatoren

Die drei Photoinitiatoren Benzoin, 2,4,6-Trimethylbenzoin und Mesitil wurden im Lösungsmittel Methanol untersucht. Die photoinduzierte Dynamik von Benzoin und 2,4,6-Trimethylbenzoin ist dabei relativ ähnlich. Ihr erster angeregter Singulett-Zustand wird innerhalb von $(0{,}7 \pm 0{,}2)$ bzw. $(1{,}2 \pm 0{,}2)$ Pikosekunden depopuliert. Der Hauptkanal für die Depopulation ist Singulett-Triplett-Interkombination. Die hohe Effizienz der Interkombination kann auf die besondere elektronische Struktur der zwei Photoinitiatoren zurückgeführt werden, welche die Existenz eines mit dem angeregten Singulett-Zustand nahezu isoenergetischen Triplett-Zustands erlaubt. Aus dem niedrigsten Triplett-Zustand erfolgt Bildung von Radikalen mit Zeitkonstanten von $(14{,}3 \pm 0{,}9)$ bzw. $(15{,}9 \pm 0{,}4)$ Pikosekunden durch α-Spaltung.

Im Falle von Mesitil wird der angeregte Singulett-Zustand dagegen deutlich schneller $((0{,}2 \pm 0{,}2)$ Pikosekunden) depopuliert. Der Hauptkanal für die Depopulation ist hier innere Konversion zwischen Singulett-Zuständen. Dies kann mit der elektronischen Struktur des Moleküls erklärt werden, welche sich deutlich von denen der anderen beiden Photoinitiatoren unterscheidet. Interkombination ist hier ein Depopulationskanal mit geringfügiger Quantenausbeute. α-Spaltung aus dem Triplett-Zustand findet innerhalb von (25 ± 2) Pikosekunden statt.

Eine Abschätzung relativer Quantenausbeuten für die Singulett-Triplett-Interkombination reproduziert frühere Ergebnisse aus der Untersuchung relativer Einbauverhältnisse der Initiator-Radikale in Polymere erstaunlich genau. Dadurch kann gezeigt werden, daß die Gesamteffizienz der photoinduzierten Radikal-Startreaktion durch Interkombination im Photoinitiator bestimmt wird, welche innerhalb der ersten Pikosekunden nach Photoanregung stattfindet.

Photoinduzierte Dynamik von fluoreszierenden Photoinitiatoren

Die photoinduzierte Dynamik der zwei Triplett-Photoinitiatoren 7-Diethylamino-3-thenoylcumarin und Isopropylthioxanthon, welche zusätzlich Fluoreszenz mit erheblichen Quantenausbeuten zeigen, wurde im Lösungsmittel Ethanol untersucht. Die Lebensdauer des ersten angeregten Singulett-Zustands von 7-Diethylamino-3-thenoylcumarin liegt bei (99 ± 1) Pikosekunden. Er wird hauptsächlich durch Fluoreszenz und Interkombination depopuliert. Die Lebensdauer des niedrigsten Triplett-Zustands liegt dabei außerhalb des untersuchten Zeitfensters. Es kann gezeigt werden, daß Bestrahlung von Molekülen im angeregten Singulett-Zustand mit Licht einer Wellenlänge von 532 nm hauptsächlich zu Abregung der Moleküle durch stimulierte Emission führt. Deshalb ist der Photoinitiator hochgradig geeignet für die Verwendung im Rahmen von Lithographie mit stimuliertem Emissionslöschen (stimulated emission depletion (STED)-Lithographie).

Isopropylthioxanthon weist dagegen eine deutlich höhere Lebensdauer des ersten angeregten Singulett-Zustands auf ((2,3 ± 0,2) Nanosekunden). Im allgemeinen würde ihn dies geeigneter für STED-Lithographie machen. Allerdings kann gezeigt werden, daß Bestrahlung des Moleküls in nahezu dem gesamten sichtbaren Spektralbereich hauptsächlich zu Lichtabsorption durch den angeregten Zustand anstatt stimulierter Emission führt. Da Lichtabsorption im angeregten Zustand zur Erhöhung der Energie des Systems anstatt zur Erniedrigung führt, ist der Prozess kontraproduktiv für STED-Lithographie. Deshalb ist das Molekül weniger geeignet für diese spezielle Anwendung als 7-Diethylamino-3-thenoylcumarin.

Contents

Contents

List of abbreviations

AO	atomic orbital
B3LYP	hybrid exchange-correlation functional consisting of B88 and LYP
BP86	exchange-correlation functional consisting of B88 VWN and P86
B88	exchange functional by Becke
BBO	β-barium borate
Bz	benzoin
c.c.	complex conjugated
CASSCF	complete active space SCF
CASPT2	CASSCF with second-order perturbational theory
CC	coupled cluster theory
CC2	second-order approximate CC
CCSD	CC with connected singles and doubles
CI	configuration interaction
CISD	CI with single and double excitations
COSMO	conductor-like screening model
CPA	chirped pulse amplification
CPA 2210	employed fs laser system
CSF	configuration state function
CT	charge-transfer
ΔOD	relative optical density
DAS	decay associated spectrum
DFM	difference frequency mixing
DFT	density functional theory

DETC	7-diethylamino-3-thenoylcoumarin
ESA	excited state absorption
FCI	full CI
FS	fused silica
fs	femtosecond
FWHM	full width at half maximum
GDD	group-delay dispersion
GGA	generalized gradient approximation
GVD	group-velocity dispersion
GVM	group-velocity mismatch
HF	Hartree-Fock theory
HOMO	highest occupied molecular orbital
I	(transmitted) intensity
I_0	irradiated intensity
IC	internal conversion
IP	ionization potential
ISC	intersystem crossing
ITX	isopropylthioxanthone
KTP	potassium titanyl phosphate
λ_p	central wavelength of the pump pulse
λ_{pr}	central wavelength of the probe pulse
LDA	local density approximation
LYP	correlation functional by Lee, Yang and Par
LUMO	lowest unoccupied molecular orbital
Me	mesitil
μs	microsecond
MO	molecular orbital
MP2	second-order Møller-Plesset perturbation theory
Nd:YAG	Neodym-doped yttrium aluminum garnet
NIR	near infrared region of the electromagnetic spectrum (800 – 3000 nm)

NOPA	noncollinear optical parametric amplifier
nm	nanometer
ns	nanosecond
OD	optical density
P86	correlation functional by Perdew
PD	photodiode
PES	potential energy surface
PLP	pulsed laser polymerization
ps	picosecond
RGA	regenerative amplifier
S_0	electronic singlet ground state
S_1	first excited singlet state
S_n	nth excited singlet state
SA	state-averaged
SCF	self-consistent field theory
SEC/ESI-MS	size exclusion chromatography / electrospray ionization mass spectrometry
SEM	stimulated emission
SHG	second harmonic generation
SFM	sum frequency mixing
SPM	self-phase modulation
SF10	dense flint glass
STED	stimulated emission depletion
TA	transient absorption
T_1	triplet ground state
T_n	nth triplet state
TDDFT	time-dependent DFT
Ti:Sa	titanium-doped sapphire
TMB	2,4,6-trimethylbenzoin
TRPE	time-resolved photoelectron

UV	ultra-violet region of the electromagnetic spectrum (200 – 400 nm)
Vis	visible region of the electromagnetic spectrum (400 – 800 nm)
VWN	correlation functional by Vosko, Wilk and Nusair

1. Introduction

Radicals are important species in many fields of chemistry and biology.[1–4] They can be defined as molecules or ions with at least one unpaired electron. Radicals naturally occur in the gas phase, namely in combustion[2,3] and atmospheric chemistry,[3] as well as in solution in biochemistry.[4] Moreover, they are important intermediates in synthetic organic and macromolecular chemistry.[1] In the latter field large-scale application of radical reactions also takes place.[5] Their discussion in the following is limited to nonmetallic radical species with one unpaired electron.

Radicals can be generated by a homolytic bond dissociation or single electron transfer. Typical dissociation energies of covalent bonds are in the range of several hundred kJ/mol. The binding energy of the H_2 molecule is, for example, 436 kJ/mol; the energy of a $C - C$ bond is in the range of 343 kJ/mol; the energy of a $C - Cl$ bond 339 kJ/mol and the energy of a $C - Br$ bond 284 kJ/mol.[6] To estimate the temperature at which such a bond thermally breaks, one can compare it to the thermal energy $k_B T$. The resulting temperatures are several thousand K. Such conditions are only achieved in the gas phase during combustion and would be destructive for most chemical structures in solution, especially sensitive biochemical systems. Thus, thermal generation of radicals under mild conditions can only be conducted with the help of precursors with extremely weak covalent bonds. Examples are the radical initiators azobisisobutyronitrile and dibenzoyl peroxide.

If a molecule instead absorbs an ultraviolet (UV) photon of 350 nm (342 kJ/mol), it is basically supplied with enough energy to break a $C - C$ single bond or a carbon – halogen bond, provided that the photoenergy is

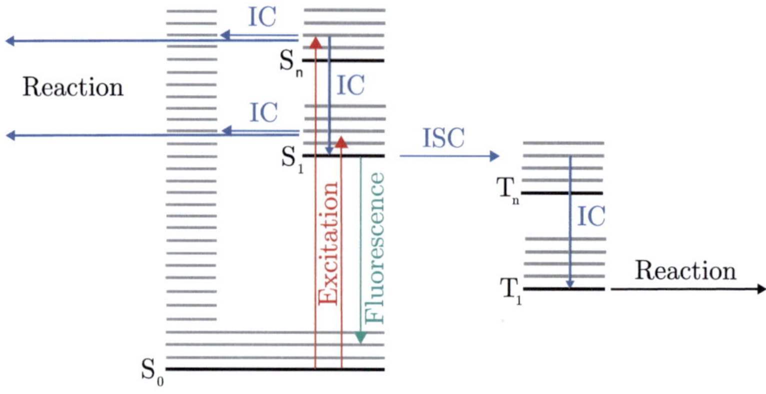

Figure 1.1.: Jabłoński-type diagram of the possible processes following excitation of a molecule from its singlet ground state (S_0) into the first (S_1) or a higher excited singlet state (S_n). These include radiationless internal conversion (IC) accompanied by vibrational relaxation between excited singlet states, into states of reaction products, or S_0 and intersystem crossing (ISC) into a triplet state. Additionally, the molecule can be deactivated by fluorescence. In addition to possible direct reaction channels from S_1 or S_n, also reaction from T_1 can take place.

selectively transferred into the bond breaking reaction. The reaction has to be faster than thermal equilibration processes, i.e. statistical distribution of the photoenergy over the molecular degrees of freedom. If these conditions are fulfilled, photogeneration of radicals can be conducted under relatively mild conditions and with high temporal and spacial control.[7,8]

The processes following light absorption in a molecule that undergoes photolysis are best discussed in the framework of a Jabłoński-type diagram as shown in Fig. 1.1: After excitation from a (in most cases) singlet ground state (S_0) into the first (S_1) or a higher excited singlet state (S_n) there are several processes competing in the depopulation of the initially excited electronic and vibrational state. One of these processes is radiationless internal conversion (IC) either into a lower excited state, S_0, or the electronic state of a reaction product. A second possible process is radiationless intersystem crossing (ISC) into the triplet manifold (T_n), followed by rapid IC into the lowest triplet state (T_1). All these processes are accompanied by vibrational relaxation within the electronic states. From T_1 reaction to products is possible. Additionally, after vibrational relaxation to the lowest vibrational state of S_1 (Kasha's rule[9]) the molecule can relax by fluorescence. Since by fluorescence the major part of the absorbed photoenergy is reemitted, the molecule is thereby in most cases brought back to the ground state without the possibility of a reaction. A further radiative process, phosphorescence from T_1 into S_0, is not included in Fig. 1.1, since it is too slow – on the order of microseconds (μs)[10] – to be important for the present work. Thus, photolytic generation of radicals takes place either directly from an excited singlet state via IC or from the lowest triplet state.

Photogenerated radicals occur both in the gas phase and solution. In gas phase one of the most important species of atmospheric chemistry, the OH radical, is partly formed via photolysis of nitrous acid.[11] It is strongly involved in the decomposition chemistry of atmospheric pollutants.[12] The chlorine radical generated by photolysis of anthropogenic chorofluorocarbons has been attracting a lot of attention, since it catalyzes the reaction of O_3 to O_2 and thereby contributes to the decline of the ozone layer.[3]

However, the investigations of the present work are focused on the mechanisms of generation and reactions of radicals in solution. In contrast to the gas phase, where radical lifetimes are basically controlled by collision numbers,[13] in solution they are in a quasi-continuous contact with surrounding solvent molecules. Thus, their lifetimes are strongly dependent not only on their own chemical nature, but also on the nature of their direct environment. Accordingly, radical lifetimes in solution span from femtoseconds (fs)[14] to more or less infinity, since persistent species like the Gomberg radical[15] have also been observed. Hence, in addition to investigations in the gas phase, interactions of the generated radicals with the solvent can be observed. Furthermore, radical generation from precursors featuring vapor pressures too low for investigations in the gas phase can be studied.

The presence of a solvent already has serious consequences a few picoseconds (ps) after photolytical formation of radicals. By photoinduced bond dissociation they are formed as geminate pairs. In the gas phase the formation dynamics normally provide them with enough kinetic energy for a ballistic separation. In solution, in contrast, they are surrounded by a joint cage of solvent molecules, which can absorb excess kinetic energy and keeps the geminate radical pair together. Recombination reactions of such radical pairs usually do not posses any energy barrier. Thus, geminate recombination is a process with high probability. The resulting influence on radical reactions is known as the solvent cage effect.[16]

A competing process is diffusional motion out of the joint solvent cage. Since energy barriers of radical reactions with other reactants are in many cases very low, their rates are commonly diffusion controlled.[3] Hence, the reaction rate is determined by the probability of the two reactants meeting during statistical, diffusive motion. Once they are close to each other, the solvent keeps them together for a high number of reactive encounters. Therefore, the probability of one of these encounters being successful is very high. Models like the Smoluchowski theory[16, 17] have been developed to quantitatively describe this behavior. However, as will be shown in Chap. 4, such a description cannot be generalized to all kinds of systems.

So far, it was assumed that the solvent does not play the role of a reactant. The situation changes if radicals are generated e.g. in solutions of monomers, which exhibit low energy barriers towards a reaction with the radical. Often, as a result of the reaction, the radical character is transferred to the monomer molecule giving it the possibility of a chain reaction with additional monomer molecules to form a polymer. The growth of the polymer chain is commonly described by random walk theory, which has – like diffusion – the general character of a statistical description, but can be constrained not to enter the same volume unit more than once.[5]

The photoreactions discussed in the present work, which all eventually lead to the generation of radicals, nicely illustrate the above discussed competition of IC, fluorescence and ISC for excited state population. The photoinduced dynamics take place on an extensive range of time scales from fs to nanoseconds (ns). Therefore, they are sorted into three chapters according to the excited state lifetimes of the precursors. The precursors described in Chap. 4 are halogenated cyclopentadienes undergoing photoinduced carbon – halogen bond dissociation. They can be regarded as polyene model systems and exhibit the shortest excited state lifetimes of < 100 fs. Here, not only the excited states dynamics are investigated, but also the subsequent dynamics in solution, which give new aspects to the microscopic picture of halogen radical solvation.

The species considered in Chap. 5 are photoinitiators, which decompose by α-cleavage of a $C - C$ bond from the triplet state. They exhibit excited state lifetimes on the order of ps. With these systems the focus lies on mechanistic aspects deciding their efficiency to induce polymerization. ISC can be identified here to be the crucial process in excited states dynamics.

The molecules investigated in Chap. 6 are non-standard triplet photoinitiators specially suitable for stimulated emission depletion (STED) lithography of structures beyond the Abbe diffraction limit.[18] Hence, they exhibit excited state lifetimes of ns and considerable fluorescence quantum yields. It can be shown that the suitability towards STED lithography is not trivially dependent on the fluorescence quantum yield, but on the interplay of the excited singlet state lifetime and the radiative coupling of the excited singlet state to higher excited states and the ground state.

For investigation of the excited state dynamics a detection method with femtosecond time resolution has to be employed. The invention of the laser by Maiman in 1960[19] was a breakthrough for the development of powerful spectroscopic techniques. Today, there are only few analytical methods in chemistry and physics which do not benefit from lasers.[20] Already very early after the invention of the laser efforts started towards generation of short laser pulses.[21] With these light sources spectroscopic techniques providing previously unmatched time resolution became possible. The first femtosecond laser pulses were generated in 1975 by a dye laser[22] constituting the field of femtosecond spectroscopy. Since then a plethora of femtosecond time-resolved spectroscopy methods have been developed[23] by many researches including A. Zewail, who was awarded the Nobel prize in 1999[24] for his achievements in this field.

The method which is primarily employed for the time-resolved investigations in the present work is femtosecond transient absorption (TA) spectroscopy. In contrast to femtosecond time-resolved photoelectron (TRPE) spectroscopy, which was additionally used for investigating halogenated cyclopentadienes, it is applicable to samples in solution and furthermore sensitive to radiative coupling between the excited electronic state and the ground state. For excitation of the samples in the ultra-violet (UV) spectral range an optical setup was built, which allows generation of femtosecond laser pulses tunable between 300 and 370 nm. The optical fundamentals of the UV light generation and the transient absorption experiment are discussed in Chap. 2.1, its experimental realization in Chap. 3.

The experimental results are supplemented by quantum chemical calculations. In Chap. 2.2 an overview of suitable quantum chemical methods to describe excited states is given with emphasis on the actually employed ones. The parameters of the calculations are given in Chap. 3.6.

2. Fundamentals

This chapter is divided into two sections. In Sect. 2.1 the fundamentals of femtosecond time-resolved spectroscopy are examined. It starts with a discussion of the properties of femtosecond laser pulses and their interactions with transparent media, proceeds with describing the generation of the pulses and ends with inspection of the pump-probe technique and the transient absorption (TA) spectroscopy method. In Sect. 2.2 an overview of theoretical methods to describe molecules in their electronically excited states is given with focus on the methods, which are employed in the present work.

2.1. Femtosecond time-resolved spectroscopy

Here, the fundamentals of the employed femtosecond time-resolved spectroscopy methods shall be discussed briefly. For a general overview of laser light generation and its special properties the reader is referred e.g. to Ref. 25, for overviews of the historical evolution of femtosecond spectroscopy to Refs. 24 and 26.

2.1.1. Interactions between femtosecond pulses and transparent media

The discussion of the properties of femtosecond laser pulses and their interaction with transparent media is focused here on a number of aspects, which are important for the performed experiments. For general literature on this topic the reader is referred to Refs. 27–31.

Dispersion

The time profiles of femtosecond laser pulses are in the region of a few harmonic cycles of the electromagnetic field. A Gaussian-shaped pulse with a full width at half maximum (FWHM) of 30 fs and a central wavelength of 600 nm consists only of approximately 15 optical cycles. Therefore, the temporal profile of such a pulse is connected to its spectral profile by a Fourier relation, which can be regarded as the energy-time analogue to the Heisenberg uncertainty-relation. It is also known as time-bandwidth product

$$\Delta t \Delta v \geq K, \tag{2.1}$$

where Δv [Hz] and Δt [s] are the spectral and temporal width of the pulse, respectively. The value of K is dependent on the pulse shape[28] (K = 0.441 for a Gaussian shape). A pulse, for which the time-bandwidth product is equal to K, is termed as Fourier-limited. Its central frequency is time-independent, i.e. all spectral components of the pulse are equally represented within the whole pulse duration. The femtosecond pulse mentioned above, therefore, has a bandwidth of 490 cm^{-1} resulting in a pulse spectrum with a FWHM ranging from 591 to 609 nm.

Such a pulse experiences wavelength-dependent refraction when propagating through a transparent, dispersive medium. The wavelength-dependent refractive index $n(\lambda)$ of a medium is connected to the velocity of light propagation $c(\lambda)$ in the medium according to the following relation[30]

$$n(\lambda) = \frac{c_0}{c(\lambda)}, \tag{2.2}$$

where c_0 is the vacuum speed of light. The effect is commonly observed e.g. in rainbows, where the solar spectrum is dispersed into its components by refraction in atmospheric water droplets.

Two effects originating from dispersion can be distinguished.[29] The first, group-delay dispersion (GDD), describes the overall retardation of the pulse dependent on its central wavelength. It has to be taken into account when two pulses of different central wavelengths interact while propagating through a dispersive medium, e.g. a liquid sample. The relative delay of one pulse with respect to the other by GDD is termed as group-velocity mismatch (GVM).

The second effect, group-velocity dispersion (GVD), is based on the different propagation velocities of each spectral component of the broadband femtosecond pulse. Thus, a Fourier-limited pulse is temporally broadened while traveling through a dispersive medium, and also its central frequency shifts within its time profile. The pulse is therefore also termed as "chirped". The same effect can however also be utilized to compensate for a chirp, if the wavelength-dependence of the refractive index is chosen to counteract the chirp.

Second-order nonlinear optics

When light propagates through a transparent medium, its electric field $\vec{E}(t)$ induces a dielectric polarization $\vec{P}(t)$ in the medium by deflection of its electrons. In most cases the response of the medium to the electric field is linear. However, for a correct description of the interaction in the presence of high field intensities the vector components of $\vec{P}(t)$ have to be expanded into a series[27,29] according to

$$
\begin{aligned}
P_i(t) &= \varepsilon_0 \sum_j \chi_{ij}^{(1)} E_j(t) + \varepsilon_0 \sum_{jk} \chi_{ijk}^{(2)} E_j(t) E_k(t) \\
&\quad + \varepsilon_0 \sum_{jkl} \chi_{ijkl}^{(3)} E_j(t) E_k(t) E_l(t) + \cdots \\
&= P_i^{(1)}(t) + P_i^{(2)}(t) + P_i^{(3)}(t) + \cdots .
\end{aligned}
\tag{2.3}
$$

Here, ε_0 is the vacuum permittivity and $\chi_{ij}^{(1)}$, $\chi_{ijk}^{(2)}$, and $\chi_{ijkl}^{(3)}$ are tensor components of the linear, second-order, and third-order nonlinear optical susceptibility. $\chi^{(1)}$ is a second-rank tensor, $\chi^{(2)}$ a third-rank tensor and so forth. The indices i,j,k,l refer to Cartesian coordinates. Nonlinear optical phenomena can, therefore, be likewise categorized with respect to their order of nonlinearity.

It can be estimated that the amplitude of the electric field has to approach the characteristic atomic field strength[27]

$$E_{at} = \frac{e}{4\pi\varepsilon_0 a_0} = 5.14 \cdot 10^{11} \frac{V}{m} \tag{2.4}$$

for observability of second-order nonlinear effects, where e is the elementary charge and a_0 the Bohr radius. For the observability of third-order nonlinear effects the field amplitude has to approach E_{at}^2. The amplitudes of the electromagnetic field are, in general, by far higher for pulsed lasers than for continuous-wave lasers at the same averaged output power, since the same amount of energy is not emitted continuously but in temporally confined packages. For comparison, the employed femtosecond laser system (see below) features an averaged output power of 1.7 W. Since its repetition rate is 1 kHz, this corresponds to an energy of 1.7 mJ/pulse. The averaged light intensity within the pulse duration of 160 fs FWHM is, therefore, I = $1.0 \cdot 10^{10}$ W/cm^2 assuming an effective beam area of 1.1 cm^2 with a homogeneous intensity distribution. Hence, the electric field strength is $7.5 \cdot 10^{14}$ V/m according to

$$E = \sqrt{\frac{2I}{\varepsilon_0 c}} \tag{2.5}$$

and thus well in the regime of second-order nonlinearities.[29]

Likewise, most effects important for understanding the frequency conversion techniques employed in the experimental setup, which will be discussed in Chap. 3, refer to the second-order nonlinear response of transparent media

$$P_i^{(2)}(t) = \varepsilon_0 \sum_{jk} \chi_{ijk}^{(2)} E_j(t) E_k(t). \tag{2.6}$$

The effects of second-order nonlinearities can be best explained assuming the electric field to be fully described by a harmonic oscillation with a single frequency ω in a single direction x (i.e. linearly polarized, monochromatic light)

$$E_x(t) = E_x e^{-i\omega t} + E_x^* e^{i\omega t} = \vec{E} e^{-i\omega t} + c.c., \tag{2.7}$$

where c.c. stands for complex conjugated. Simultaneously, Eq. 2.6 is reduced to

$$P_i^{(2)}(t) = \varepsilon_0 \chi_{ixx}^{(2)} E_x(t) E_x(t). \tag{2.8}$$

Inserting Eq. 2.7, one yields

$$P_i^{(2)} = 2\varepsilon_0 \chi_{ixx}^{(2)} E_x E_x^* + \left[\varepsilon_0 \chi_{ixx}^{(2)} E_x^2 e^{-2i\omega t} + c.c. \right]. \tag{2.9}$$

Thus, the second-order nonlinear polarization in this case consists of a static contribution, the so-called optical rectification, and a contribution oscillating with the double frequency of the driving field.[27] The latter leads to emission of light with the double frequency of the incident light. Consequently, the process is called second harmonic generation (SHG).

Assuming the electric field of the incident light consisting of two components oscillating at two different frequencies ω_1 and ω_2 but again only in x-direction

$$E_x(t) = E_{1,x} e^{-i\omega_1 t} + E_{2,x} e^{-i\omega_2 t} + c.c., \tag{2.10}$$

the same considerations as above lead to an expression of the second-order nonlinear polarization[27]

$$P_i^{(2)}(t) = \varepsilon_0 \chi_{ixx}^{(2)} \left[\underbrace{E_{1,x}^2 e^{-2i\omega_1 t}}_{\text{SHG } \omega_1} + \underbrace{E_{2,x}^2 e^{-2i\omega_2 t}}_{\text{SHG } \omega_2} \right.$$
$$+ \underbrace{2 E_{1,x} E_{2,x} e^{-i(\omega_1 + \omega_2)t}}_{\text{SFM}} + \underbrace{2 E_{1,x} E_{2,x}^* e^{-i(\omega_1 - \omega_2)t}}_{\text{DFM}} + c.c. \left. \right]$$
$$+ 2\varepsilon_0 \chi_{ixx}^{(2)} \left[E_{1,x} E_{1,x}^* + E_{2,x} E_{2,x}^* \right]. \tag{2.11}$$

The first two terms in the bracket are already introduced and belong to SHG at the frequencies ω_1 and ω_2. The third term refers to an oscillation of the polarization and therewith light generation at a frequency, which represents the sum of the two frequencies of the driving electric field. Hence, the process is known as sum frequency mixing (SFM). The last term represents generation of light at the difference frequency of the two components of the driving field and is called difference frequency mixing (DFM). In this case it is assumed that $\omega_1 > \omega_2$.

SHG, SFM and DFM can also be described in the particle picture. In the case of SHG two photons of the same energy (ω) merge to generate a photon of the double energy (2ω). In SFM the same process takes place with two photons of different energy (ω_1 and ω_2) merging to a photon of $\omega_1 + \omega_2$. All processes are restricted to energy conservation. Thus, for DFM the interaction of two photons of different energy has to divide the photon of ω_1 in one photon of $\omega_1 - \omega_2$ and one additional photon of ω_2. Hence, the process is also known as optical parametric amplification, since light of the lower frequency is amplified by light of the higher frequency.[27]

Whether, and which, one of the above described processes is observed, depends on the nature of the transparent medium and the geometry of inter-action. Many media exhibit large nonlinear optical susceptibilities. However, the occurrence of non-zero $\chi^{(2)}_{ijk}$ tensor components is connected to the symmetry of the medium's unit cell. In media with a centrosymmetric unit cell the anharmonicity for the oscillating electrons is symmetric and therefore all second-order contributions to the polarization vanish.[27] Thus, second-order nonlinearities are only observed in media with a non-centrosymmetric unit cell, where the anharmonicity is unsymmetric.

Such media, for example β-barium borate (BBO), exhibit an optical axis and are, therefore, optically anisotropic.[27] This property is also expressed by the effect of birefringence, a splitting of unpolarized light into perpendicularly polarized components via a refractive index, which is dependent on the polarization plane. The geometrical implications are best visualized

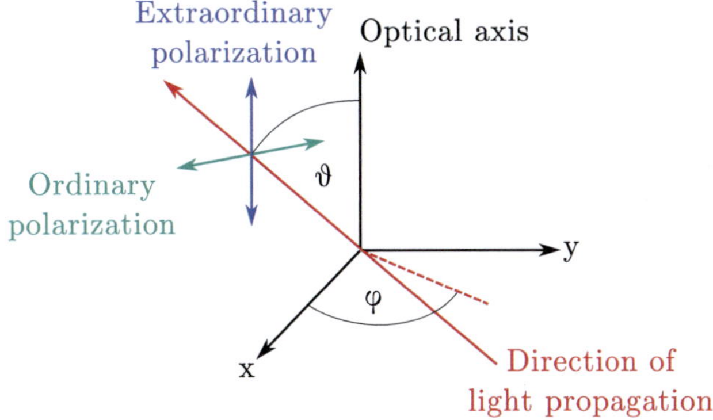

Figure 2.1.: Scheme visualizing the geometrical aspects of light propagating through a birefringent crystal. The direction of light propagation (red) can be described relative to the crystal unit cell (black) by its angle ϑ with respect to the optical axis of the crystal and the angle φ between its projection on the xy-plane (dotted) and the x-axis. The ordinary polarization plane of the light is perpendicular to the plane spanned by the direction of propagation and the optical axis, the extraordinary polarization plane parallel.

with the help of Fig. 2.1: The relative orientation of the direction of light propagation with respect to the crystal unit cell can be described by the angle ϑ between the propagation direction and the optical axis and the angle φ between the projection of the propagation direction on the xy-plane and the x-axis. The polarization plane perpendicular to the plane spanned by the direction of propagation and the optical axis is defined as the ordinary polarization plane, since its refractive index is not affected by the relative orientation of the light beam to the optical axis. This plane includes the x-direction of the oscillations in the examples discussed above. The polarization plane perpendicular to the ordinary is defined as the extraordinary polarization plane. This plane includes the i-direction of the examples above. Its refractive index is dependent on the angle ϑ according to

$$\frac{1}{n_e^2(\vartheta)} = \frac{\sin^2\vartheta}{n_e^2(\vartheta = 90°)} + \frac{\cos^2\vartheta}{n_o^2}, \qquad (2.12)$$

where $n_e(\vartheta)$ is the ϑ-dependent extraordinary refractive index, and $n_o = n_e(\vartheta = 0°)$ the ordinary refractive index.[27] Since the refractive index is also wavelength-dependent (i.e. also frequency-dependent) and connected to the velocity of light propagation in the specific medium according to Eq. 2.2, the driving electric field of e.g. a SHG process and the second harmonic normally propagate with different velocities through the medium. The generated second harmonics therefore vanish by destructive interference and no SHG can be macroscopically observed. However, by tuning the angle ϑ, the ordinary (x-direction) and extraordinary (i-direction) refractive indices and therewith the propagation velocities of the driving field and the second harmonic can be matched. This is the so-called type I phase-matching condition.[27] It can be selectively fulfilled for each process shown in Eq. 2.11 by changing ϑ.

The selectivity of phase-matching is not fully given in the case of femtosecond pulses (see the description in Chap. 3.2). The reasons can be ascribed to the spectral width of femtosecond pulses. To achieve broadband pulses from e.g. SFM, the phase-matching condition has to be fulfilled for the whole spectral envelope of the generating pulses. Therefore, thin BBO crystals have to be employed, where the influences of destructive interference are not yet severe.[28] As a result, at the optimum ϑ for SFM often also SHG from ω_1 and ω_2 is observable with low efficiency. In practice BBOs are employed, which are cut at specific angles ϑ. Thus, the angle tuning of the BBO and therewith the losses by reflection at the BBO surface can be minimized.

Third-order nonlinear optics

The third-order nonlinear response of the polarization $P_i^{(3)}(t)$ is proportional to the third power of the electric field strength according to

$$P_i^{(3)}(t) = \varepsilon_0 \sum_{jkl} \chi_{ijkl}^{(3)} E_j(t) E_k(t) E_l(t). \tag{2.13}$$

The observability of third-order nonlinear effects is not limited any more to birefringent media. Within the regime of third-order nonlinearity e.g. third harmonic generation is possible. Additionally, the refractive index becomes intensity-dependent according to

$$n = n_0 + \frac{1}{2}n_2 I \tag{2.14}$$

with the intensity I, which is connected to the electric field strength according to Eq. 2.5.[29] The effect leads to self-focusing of intense laser pulses, which is known as the optical Kerr effect. Additionally, it acts on the phase of the pulse leading to self-phase modulation (SPM).[32–34] By SPM additional components are added to the pulse spectrum eventually leading to a substantial spectral broadening. Accordingly, SPM is employed to create white-light continua from fs pulses. Since SPM is a third-order effect, fs pulses normally have to be strongly focused to achieve field strengths high enough to enter the regime of third-order nonlinearity.

2.1.2. Femtosecond pulse generation

The considerations about femtosecond pulse generation and femtosecond lasers are confined here to the femtosecond laser system, which was actually used for the experiments discussed later, a CPA 2210 (Clark-MXR). For an overview of the historic development of femtosecond lasers the reader is referred to Ref. 35, for a more detailed introduction into theory and techniques of femtosecond pulse generation to Refs. 28 and 29.

Femtosecond lasers suitable for pump-probe spectroscopy normally consist of a solid state femtosecond oscillator and an amplifier employing the chirped pulse amplification (CPA) method.[28,36,37] In the CPA 2210 laser system, femtosecond pulses at a central wavelength of 1550 nm and a repetition rate of 34 MHz[38] are generated by a fiber oscillator doped with Er^{3+}, which is pumped at $\lambda = 980$ nm by a solid-state laser diode.[39–42] The output is frequency-doubled to a central wavelength of 775 nm in a potassium ti-tanyl phosphate (KTiOPO$_4$, KTP) crystal.[43]

The frequency doubled output is amplified in a CPA scheme. Thereby, at first the femtosecond pulses are substantially stretched in time, then amplified in a suitable crystal, which is pumped by ns laser pulses, and finally recompressed to a fs duration. The amplification process employed here must not be confused with parametric amplification as discussed in Sect. 2.1.1. In the present process the pulses are amplified by stimulating emission from the pumped crystal. The reasons for the CPA procedure are, on the one hand, a better amplification efficiency due to a better time overlap between the fs and the pump laser pulses in the amplifier crystal and, on the other hand, protection of the employed optics due to distinctly lower peak intensities of the amplified pulse.

In the CPA 2210 amplifier the seed pulses from the fs oscillator are brought to a length of several hundred ps by reflective dispersion gratings, which fulfill the same task as a dispersive medium (see Sect. 2.1.1). Subsequently, they are coupled into a regenerative amplifier[28, 36, 37, 44–46] (RGA). Thereby, with a system consisting of a Faraday isolator, two Glan-Taylor polarizers, and a Pockels cell, 1000 pulses/s are selected from the 34 MHz pulse train and amplified by several passes through a titanium-doped sapphire (Ti:Sa) crystal, which is pumped by a Q-switched Nd:YAG laser with a repetition rate of 1 kHz and an energy of 11 mJ/pulse. After amplification the pulses are coupled out into the compressor mainly consisting of a transmittive dispersion grating, which is passed 4 times. The emitted pulses have a central wavelength of 775 nm, a length of 160 fs FWHM, and an energy of ≈ 1.7 mJ/pulse.

2.1.3. Frequency conversion in the visible and near-infrared spectral range

A noncollinear optical parametric amplifier (NOPA),[47–49] as employed for the TA experiments, combines a number of the nonlinear optical processes discussed in Sect. 2.1.1 and is capable of producing femtosecond pulses in the visible (Vis, 470 – 750 nm) and in the near-infrared (NIR, 850 – 1600 nm) spectral range. It is typically pumped by laser pulses of 200 – 250 µJ/pulse, in the present case by a fraction of the CPA 2210 output.[50] A very small amount of the pulse energy (1 µJ) is separated and focused into a sapphire plate to produce a white light continuum via SPM spanning from 450 to 750 nm. As a result of the SPM process, it is highly chirped to a length in the range of 1 ps. The rest of the pulse energy is frequency-doubled by SHG in a BBO crystal and noncollinearly overlapped with the white light continuum in a second BBO. Due to the chirp of the continuum, selected spectral regions can be amplified via a parametric process (DFM) in the BBO, not only by changing the phase-matching angle ϑ, but also by tuning the time overlap between the ps white light continuum (ω_2) and the fs second harmonic (ω_1, see Eq. 2.11). For achieving pulse energies in the µJ range, either the amplified photons of ω_2 (Vis) or the photons of $\omega_1 - \omega_2$ (NIR) are further parametrically amplified in a second stage. The amplified pulses are prolonged with respect to the initial laser pulses due to effects of white light generation and GVD in transmittive optics and therefore far from their Fourier limit. The GVD can be compensated by a compressor similar to the one discussed in Sect. 2.1.2. In contrast to Sect. 2.1.2, the dispersion compensation is provided by two Brewster prisms of dense flint glass (SF10). With this compressor, pulse lengths of 30 fs in the visible and 50 fs in the NIR spectral regimes are easily obtainable.[48]

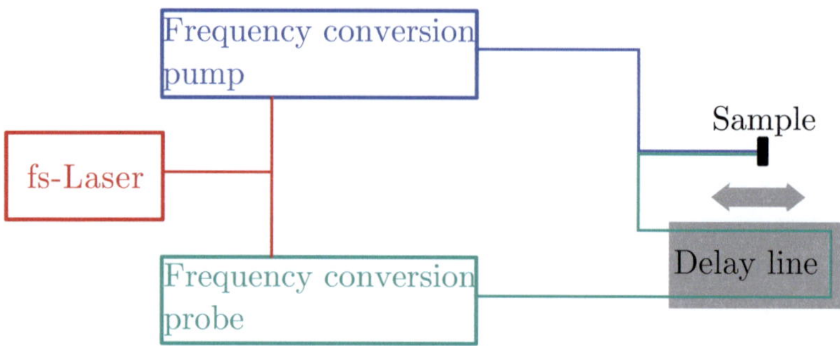

Figure 2.2.: Basic concept of fs pump-probe spectroscopy.

2.1.4. Femtosecond pump-probe spectroscopy

The basic concept, which all fs pump-probe spectroscopy techniques have in common, is visualized in Fig. 2.2: The optical path of a fs laser is split up into a path for the generation of pump pulses by frequency conversion and a second path for generation of probe pulses. Subsequently, pump and probe pulses are recombined to interact with a sample. The sample is excited by a pump pulse. The time evolution of population in the excited state, which is initially prepared by the pump pulse, is investigated by the probe pulse. For this, the probe pulses can be delayed with respect to the pump pulses via an optical delay line. The diverse pump-probe techniques only differ in the type of sample, in the way the pulses interact with the sample, and the technique to detect the interaction between the probe pulses and the sample.

2.1.5. Transient absorption (TA) spectroscopy

In femtosecond transient absorption (TA) spectroscopy the population evolution in the excited state is investigated by measuring the time-dependent change in optical density (OD)

$$\Delta OD(\tau) = OD_w(\tau) - OD_{w/o} \qquad (2.15)$$

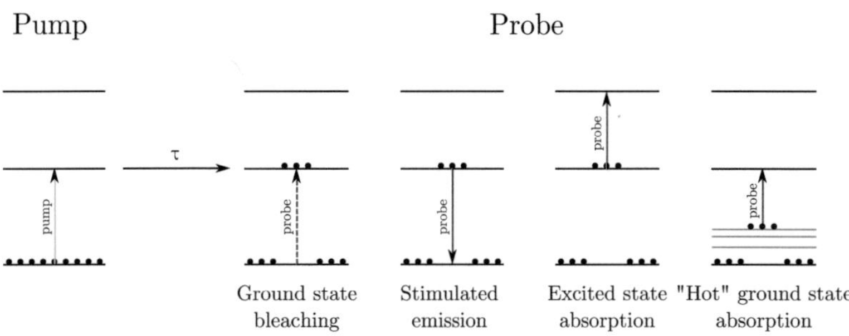

Figure 2.3.: Possible contributions to the ΔOD signal in TA spectroscopy: Ground state bleaching adds a negative contribution to ΔOD due to a reduced number of molecules in the ground state. Stimulated emission also gives a negative contribution, since the intensity of the probe beam is amplified instead of reduced due to stimulated emission of the sample. Excited state absorption contributes positive values to ΔOD due to further excitation of the sample by the probe pulse. "Hot" ground state absorption gives also a positive contribution due to the absorption of the probe pulse by vibrationally excited molecules.

of the sample at the wavelength of the probe pulse (λ_{pr}) due to the pump pulse. $OD_w(\tau)$ is the OD of the excited sample and $OD_{w/o}$ the OD of the sample without excitation. Four processes can give contributions to the observed ΔOD. They are depicted in Fig. 2.3. Excitation by the pump pulse brings a distinct amount of molecules in the excited state. If λ_{pr} is similar to the pump wavelength (λ_p), one contribution to ΔOD with a negative sign can arise from the fact that the probability to excite additional molecules by the probe pulse is reduced due to the reduced number of molecules in the ground state. This contribution is known as ground state bleaching. Another contribution to ΔOD with the same sign can result from de-excitation of the sample via emission, which is stimulated by the probe pulse (SEM). The third possible contribution with a positive sign comes from further excitation of molecules from the excited state into higher excited states, known as excited state absorption (ESA). An additional contribution can originate from molecules, which have already relaxed from an electronically excited state e.g. by IC (see the Jabłoński-type diagram in Fig. 1.1), but are highly vibrationally excited. This leads to a red-shift of the ground state absorption spectrum and therefore to a positive ΔOD contribution at a red-shifted λ_{pr}. The contribution is therefore known as "hot" ground state absorption.

Experimentally observed TA signals are often a convolution of at least two of these contributions, because they are connected to specific probabilities of how excited molecules can interact with the probe pulse. Since always an ensemble of excited molecules is under consideration, also very small probabilities are realized.

2.1.6. Time-resolved photoelectron (TRPE) spectroscopy

In time-resolved photoelectron (TRPE) spectroscopy the population evolution of the excited sample is not detected by measuring a relative OD, but by photoionizing the sample with the probe pulse and detecting the number, as well as the kinetic energy, of the photoelectrons.[51] One benefit of this method is that in contrast to TA spectroscopy the final states the molecule is ionized into can usually be identified very well. In many cases the final state is the ionic ground state, which is easily accessible by theoretical methods. One flaw of the method is that the samples have to be investigated in the gas phase. Therefore, the number of systems which can be investigated is limited to molecules with a sufficient vapor pressure. Furthermore, emission from the sample is not detectable. Obviously, the contributions shown in Fig. 2.3 do not apply to TRPE spectroscopy. Ground state bleaching is generally possible, but does not contribute to the TRPE spectra under usual experimental conditions. As mentioned above, stimulated emission is not detectable, since electrons instead of photons are detected. Excited state absorption is replaced by excited state ionization and "hot" ground state absorption is usually not detectable. The reasons for that are discussed in detail in Chap. 4.4.

2.2. Theoretical methods

The interpretation of experimental data from femtosecond spectroscopy in many cases necessitates the employment of quantum chemical methods. Since in the experiments, which are discussed in the present work, the fate of electronically excited states is investigated, an overview of different quantum chemical methods and their ability to treat electronically excited states shall be given, accompanied by a more detailed discussion of the methods which are employed here. For a general synopsis of fundamentals and methods in quantum chemistry, the reader is referred to Refs. 52–54.

Since the treatment of quantum chemical problems in most cases aims at describing the electronic ground state of molecules as accurately as possible, the time-independent Schrödinger equation

$$\hat{H}\Psi\left(\vec{r},\vec{R}\right) = E\Psi\left(\vec{r},\vec{R}\right) \tag{2.16}$$

is in most cases solved within the framework of the Born-Oppenheimer approximation,[55] where the nuclear kinetic energy is set to zero. In Eq. 2.16 $\Psi\left(\vec{r},\vec{R}\right)$ is the molecular wave function depending on nuclear ($\vec{R}$) and electronic ($\vec{r}$) coordinates, E is an energy eigenvalue of Ψ and $\hat{H}$ the molecular Hamiltonian.

The approximation allows for the separation of the Schrödinger equation and the wave function into an electronic Schrödinger equation and wave function, which is only parametrically dependent on nuclear coordinates, and a nuclear Schrödinger equation and wave function, where the eigenvalue of the electronic Schrödinger equation acts as a potential for the nuclei. The Born-Oppenheimer approximation thereby provided chemists with the very illustrative picture of purely electronic eigenstates each featured with a set of nuclear vibrational and rotational levels and the concept of nuclei moving either classically or quantum-mechanically on an electronic potential energy surface (PES).

However, a consequence of the Born-Oppenheimer approximation is that not only the coupling between nuclear and electronic states is neglected, but also the existence of additional solutions to the electronic Schrödinger equation, e.g. other electronic eigenstates,[56,57] and their coupling to the electronic state under consideration. Thus, from an accurate description of the electronic ground state, one cannot automatically describe energy differences to excited states. Nevertheless, the approximation holds very well for many areas of the ground state PES and only breaks down in the vicinity of state degeneracies, where the neglected couplings become important.

Two approaches exist for calculation of excited states, whereas the first one is based on methods developed for accurate description of the electronic ground state. In this single-referential approach, excited state properties are approximated from the ground state wave function. The advantages are obvious: well established, efficient, and reliable methods can be employed to generate the ground state wave function. There are two major disadvantages: as the excited states are not implicitly included into the wave function, systematic errors can occur dependent on the employed method. Additionally, excited states can only be calculated, when a stable ground state wave function is obtainable.

In the second, multi-referential approach a set of different electronic states is included in a single wave function, which is based on a group Born-Oppenheimer approximation.[56,57] By the implicit treatment of ground state and excited states together, all states are represented qualitatively correctly even in situations when the ground state wave function of the single-referential approach is unstable. However, multi-reference methods do not work as robustly as single-reference methods. Furthermore, their application is not as straightforward as most single-reference methods, since they require some experience by the user and a detailed knowledge about the particular problem. Additionally, the computational demand for achieving quantitative results is very high. Accordingly, to date the approach is only applicable to small systems.

2.2.1. Single-reference methods

Since the ground state is in most cases well separated from other electronic states, coupling to higher states can often be neglected. Within the Hartree-Fock (HF) (or self-consistent field (SCF)) ansatz, the wave function of the electronic ground state is prepared as an antisymmetrized product of single electron wave functions (or molecular orbitals (MOs)), a so-called Slater determinant. The solution is approximated by fractionizing the equation into a set of pseudo one-electron differential equations, the Hartree-Fock equations, which are coupled by the electron-electron potential. Applying the variational principle, the system of coupled differential equations is solved iteratively (for details see e.g. Ref. 53).

One flaw of the Hartree-Fock ansatz is the inherent approximation of the electrons as independent particles. Each electron experiences only an averaged potential of the remaining electrons, which is not mirrored in reality. Hence, the exact ground state energy can only be approximated to a distinct amount known as the HF limit.

The difference between the exact energy and the HF limit is due to correlated motion of electrons. Two contributions are commonly discerned, dynamic electron correlation, which originates from electrons avoiding each other in the same space, and static or near-degeneracy electron correlation, which becomes important when electronic states come close to each other and the Born-Oppenheimer approximation breaks down.[53] For achieving chemical accuracy, the error of the HF limit has to be overcome. For this purpose, a considerable number of so-called post-HF methods have been developed.

The exact solution of the Schrödinger equation in the basis set limit can be achieved by taking the optimized Slater determinant from an HF calculation (Φ_{HF}) and generating all possible excited Slater determinants Φ_i from it:

$$\Phi_{CI} = c_0\Phi_{HF} + \sum_i c_i\Phi_i \tag{2.17}$$

The Schrödinger equation can then be set up as a matrix eigenvalue equation

$$
\begin{pmatrix}
H_{00} - E & H_{01} & \cdots & H_{0n} \\
H_{10} & H_{11} - E & \cdots & H_{1n} \\
\vdots & \vdots & \ddots & \vdots \\
H_{n0} & H_{n1} & \cdots & H_{nn}
\end{pmatrix}
\begin{pmatrix}
c_0 \\ c_1 \\ \vdots \\ c_n
\end{pmatrix}
=
\begin{pmatrix}
0 \\ 0 \\ \vdots \\ 0
\end{pmatrix},
\qquad (2.18)
$$

where the elements of the Hamilton matrix are defined as $H_{ij} = \langle \Phi_i | \hat{H} | \Phi_j \rangle$. The ground state energy is the lowest eigenvalue of equation 2.18, the ground state wave function can be created from the corresponding eigenvector, which contains the values for the expansion coefficients c_i of equation 2.17. This approach is known as full configuration interaction (FCI). Unfortunately, due to its high computational demand, it can only be applied to the smallest chemical systems. Post-HF methods, which can be applied to the systems of interest in the present work, save computational effort by truncating the expansion in equation 2.17. Thereby, the two requirements, the accuracy of the ground state energy and the computational demand, have to be balanced out carefully. One approach is, for example, configuration interaction with single and double excitations (CISD), where the expansion is restricted to excitations of one and two electrons from the HF Slater determinant, another approach is coupled cluster (CC) theory, which is described in more detail in the following section.

Coupled cluster theory

In CC theory[53,54] an excitation operator

$$
\hat{T} = \sum_{i=1}^{N} \hat{T}_i
\qquad (2.19)
$$

is defined, where N is the number of electrons of the treated system and $\hat{T}_i$ produces all possible i-fold excited Slater determinants from Φ_{HF}, for example:

$$\hat{T}_1 \Phi_{HF} = \sum_i^{occ.} \sum_j^{unocc.} t_i^j \Phi_i^j, \tag{2.20}$$

where Φ_i^j is a Slater determinant, in which an electron i is excited to an orbital j, and t_i^j is the corresponding CC amplitude. The ansatz for the CC wave function is (cf. Eq. 2.17)

$$\Phi_{CC} = e^{\hat{T}} \Phi_{HF}. \tag{2.21}$$

and the CC Schrödinger equation is

$$\hat{H} e^{\hat{T}} \Phi_{HF} = E_{CC} e^{\hat{T}} \Phi_{HF}. \tag{2.22}$$

$e^{\hat{T}}$ can be expanded in a power series and the summands can be arranged with respect to the level of excitation they cause

$$\begin{aligned} e^{\hat{T}} &= 1 + \hat{T}_1 + \left(\hat{T}_2 + \frac{1}{2} \hat{T}_1^2 \right) + \left(\hat{T}_3 + \hat{T}_2 \hat{T}_1 + \frac{1}{6} \hat{T}_1^3 \right) \\ &+ \left(\hat{T}_4 + \hat{T}_3 \hat{T}_1 + \frac{1}{2} \hat{T}_2^2 + \frac{1}{2} \hat{T}_2 \hat{T}_1^2 + \frac{1}{24} \hat{T}_1^4 \right) + \cdots . \end{aligned} \tag{2.23}$$

It is thereby differentiated between connected cluster excitations produced by only one $\hat{T}$ operator (e.g. $\hat{T}_3$) and disconnected cluster excitations produced by a product of several $\hat{T}$ operators (e.g. $\hat{T}_2 \hat{T}_1$). Because of high computational demand the sum of Eq. 2.19 is always truncated very early. For the common CC with connected singles and doubles (CCSD) variant, it is restricted to $\hat{T} = \hat{T}_1 + \hat{T}_2$. The expansion of equation 2.23 therefore changes to

$$\begin{aligned} e^{\hat{T}_1 + \hat{T}_2} &= 1 + \hat{T}_1 + \left(\hat{T}_2 + \frac{1}{2} \hat{T}_1^2 \right) + \left(\hat{T}_2 \hat{T}_1 + \frac{1}{6} \hat{T}_1^3 \right) \\ &+ \left(\frac{1}{2} \hat{T}_2^2 + \frac{1}{2} \hat{T}_2 \hat{T}_1^2 + \frac{1}{24} \hat{T}_1^4 \right) + \cdots . \end{aligned} \tag{2.24}$$

It can be seen that, although only $\hat{T}_1$ and $\hat{T}_2$ operators are present in the expansion, also triply and higher excited Slater determinants are produced by multiple application of the operators $\hat{T}_1$ and $\hat{T}_2$. However, these higher excitations are disconnected cluster excitations and therefore their description is poorer than in the FCI case. This is the main difference to truncated CI methods like CISD, where only singly and doubly excited Slater determinants are included. The CC amplitudes and E_{CC} can then be evaluated from

$$\langle \mu | \hat{H}^T | \Phi_{HF} \rangle = 0 \tag{2.25}$$

and

$$E_{CC} = \langle \Phi_{HF} | \hat{H}^T | \Phi_{HF} \rangle, \tag{2.26}$$

respectively.[54] $\hat{H}^T = e^{-\hat{T}} \hat{H} e^{\hat{T}}$ is the similarity-transformed Hamiltonian and μ are singly and doubly excited Slater determinants.

The second-order approximate CC (CC2) method,[54,58–61] which is employed in the present work, is an approximate CCSD method. Here the Hamiltonian has the form of

$$\hat{H}(t) = \hat{F} + \hat{\phi} + \hat{V}(t) = \hat{H}_{MP} + \hat{V}(t), \tag{2.27}$$

where $\hat{H}_{MP}$ is a Hamiltonian derived from Møller-Plesset perturbation theory[53,54] consisting of the Fock operator $\hat{F}$ from HF theory and a fluctuation potential $\hat{\phi}$. In the case of CC2 an additional time-dependent one-electron perturbation $\hat{V}(t)$ is added. The CC2 amplitudes are evaluated according to

$$\langle \mu_1 | [\hat{H}^T, \hat{T}_2] + \hat{H}^T | \Phi_{HF} \rangle = 0 \tag{2.28}$$

and

$$\langle \mu_2 | [\hat{F}, \hat{T}_2] + \hat{H}^T | \Phi_{HF} \rangle = 0. \tag{2.29}$$

The CC2 energy can be evaluated by

$$E_{CC2} = \left\langle \Phi_{HF} \left| \hat{\phi} \left(\hat{T}_2 + \frac{1}{2} \hat{T}_1 \hat{T}_1 \right) \right| \Phi_{HF} \right\rangle.$$ (2.30)

The resulting CC2 energies are slightly more accurate than those obtained from second-order Møller-Plesset perturbation theory (MP2).[54] But the main improvement over MP2 is the possibility to calculate excited state energies with the same accuracy.

Excited state energies and oscillator strengths, i.e. radiative couplings between ground state and excited states, are evaluated from an optimized CC2 wave function with linear response theory.[58,60–62] In this approach the electromagnetic field of light is treated as a time-dependent perturbation to the Schrödinger equation. The time dependent expectation value of the excitation operator $\hat{\mu}(t)$ can thereby be expanded with respect to the level of perturbation. The linear term exhibits poles at frequencies of the perturbation, which match the energy difference between the ground state and an excited state and therefore applies to one-photon excitations. These frequencies are also eigenvalues of the CC2 Jacobian matrix

$$A_{\mu_i \nu_j} = \begin{pmatrix} \langle \mu_1 | [(\hat{H}^T + [\hat{H}^T, \hat{T}_2]), \hat{\tau}_{\nu_1}] | \Phi_{HF} \rangle & \langle \mu_1 | [\hat{H}^T, \hat{\tau}_{\nu_2}] | \Phi_{HF} \rangle \\ \langle \mu_2 | [\hat{H}^T, \hat{\tau}_{\nu_1}] | \Phi_{HF} \rangle & \langle \mu_2 | [\hat{F}, \hat{\tau}_2] | \Phi_{HF} \rangle \end{pmatrix},$$ (2.31)

which is the stability matrix of the CC wave function with respect to the cluster amplitudes.[61] The ν_i are excited Slater determinants different from the μ_i. It has to be mentioned that the high accuracy for calculation of excitation energies is only maintained in the case of excited states, for which one singly excited Slater determinant is a good zeroth-order approximation.[61] Additionally, CC2 results are not reliable in the vicinity of a conical intersection, especially if the reference state (ground state) is involved.

Density functional theory

One in principle totally different approach to ground state energies, density functional theory (DFT), is based on the Hohenberg-Kohn theorem, which states that the electronic ground state energy is a functional of the electron density ρ.[63] However, the Kohn-Sham approach, which is the only practically relevant one for evaluation of molecular ground state energies, also employs the optimization of single electron wave functions, the Kohn-Sham orbitals.[64] Here – as in the HF approach – the electrons are approximated as independent particles and furthermore as non-interacting. For this case the energy as a functional of the electron density can be exactly evaluated using the same mathematical methods as HF (for details see e.g. Ref. 53). The result is then corrected by the exchange-correlation functional $E_{XC}[\rho]$. In many cases it is additionally separated into a pure exchange functional $E_X[\rho]$ and a pure correlation functional $E_C[\rho]$, which both can be approximated by different methods. Examples are the local density approximation (LDA) and the generalized gradient approximation (GGA). The first approach is employed, for example, in the correlation functional of Vosko, Wilk and Nusair (VWN),[65] the latter in the exchange functional proposed by Becke (B88)[66] and in the correlation functionals proposed by Lee, Yang and Par (LYP)[67] and by Perdew (P86).[68]

The BP86 exchange-correlation functional employed for ground state geometry optimizations in the present work consists of the B88 exchange and the VWN and the P86 correlation functionals. The very popular B3LYP functional,[69,70] which is employed here to calculate ground state energies and for TDDFT calculations (see below), is a so-called hybrid functional and employs a weighted mixture of the VWN, B88 and LYP functionals together with the exact exchange energy derived from HF.

Excitation energies are calculated employing a linear response theory approach similar to that discussed above[71,72] and is known as time-dependent DFT (TDDFT). In this case the time-dependent perturbational theoretical treatment is applied to the Kohn-Sham density matrix of a DFT calculation. TDDFT is known to evaluate excitation energies to states with singly excited character with an unmatched combination of accuracy and low computational demand,[73] allowing for the calculation of electronic absorption spectra of large molecules like the photoinitiators employed in the present work. However, the method is known to poorly handle states with doubly excited character. Furthermore, as CC2, TDDFT fails to describe PESs in the vicinity of a state crossing.[74]

2.2.2. Multi-reference methods

For the generation of multi-reference wave functions an approach similar to FCI (see above) can be made. At a molecular geometry, where a single Slater determinant is a good approximation to the ground state, a HF Slater determinant is generated. As in CI, the Slater determinant is then expanded in a series including excited Slater determinants or configuration state functions (CSFs). Employing a matrix eigenvalue equation as Eq. 2.18, the expansion coefficients and – as opposed to CI – the single electron wave functions from the HF Slater determinant are optimized with respect to a set of eigenstates.[53] As for FCI the number of CSFs has to be limited.

State averaged complete active space SCF

Within complete active space SCF (CASSCF) theory, a CAS is constituted from a limited number of occupied and unoccupied MOs of the reference HF Slater determinant. All possible distributions of the CAS electrons to the CAS MOs are included as CSFs in the expansion of the wave function and subsequently optimized with respect to a set of electronic states. The CAS MOs have to be selected carefully to make sure that all electronic states are well described by a linear combination of the CSFs produced from them.[75] A converged state-averaged (SA) CASSCF wave function accounts very well for near-degeneracy electron correlation and, therefore, treats electronic states qualitatively correctly even at points of degeneracy.[75] However, the computational demand of the method rapidly increases with an increasing CAS, and the convergence becomes more and more difficult with an increasing number of simultaneously averaged states. Additionally, the treatment of dynamic electron correlation is as poor as in HF. Like in the case of HF, there exist methods based on the CASSCF wave function, which can account for dynamic electron correlation. Examples are multi-reference CISD[76] and multi-reference CC,[77] which are computationally demanding and therefore can be only applied to small systems. An alternative approach is complete active space SCF with second-order perturbational theory (CASPT2),[78, 79] which exhibits a broader applicability.

3. Experiments

The experimental methods applied in the present work are primarily steady state UV-Vis absorption spectroscopy in solution and femtosecond time-resolved transient absorption (TA) spectroscopy in solution. Additional methods, namely femtosecond time-resolved photoelectron (TRPE) spectroscopy and fluorescence spectroscopy, which were employed in collaboration with other groups, are briefly discussed. For details on these methods the reader is referred to the literature which is cited in connection with the description of the experimental conditions.

3.1. Steady-state UV-Vis absorption spectroscopy

For steady-state UV-Vis absorption spectroscopy in solution a Cary 500 UV-Vis-NIR spectrometer (Varian) was used. All spectra were obtained in fused-silica (FS, Hellma Analytics) cuvettes with an optical path length of 1 mm and normalized to the concentration of the sample according to the Lambert-Beer law

$$\varepsilon(\lambda) \left[\frac{l}{mol \cdot cm} \right] = \frac{OD}{c \cdot d},$$
(3.1)

where $\varepsilon(\lambda)$ is the wavelength-dependent molar decadic extinction coefficient, OD the optical density, c [mol/l] the concentration, and d [cm] the optical path-length of the sample.

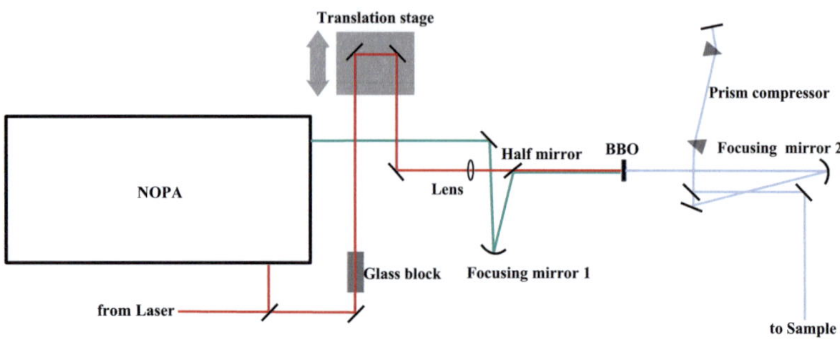

Figure 3.1.: Schematic picture of the UV-NOPA, which was set up for the experiments. The beam path of the 775 nm pulses is colored in red, the path of the NOPA output in green and the path of the UV pulses in blue.

3.2. Frequency conversion in the UV spectral range

For generation of femtosecond pulses tunable between 300 and 370 nm a UV-NOPA, which is depicted in Fig. 3.1, was built according to Ref. 80. The system includes a commercial NOPA system (Clark MXR), which is pumped by 250 µJ of the laser output from the CPA 2210 (see Chap. 2.1.2) and operated in the visible spectral range. The output energy is between 5 and 10 µJ/pulse, depending on the central wavelength. An additional fraction (100 µJ) of the laser output is sent over a delay line including a manual translation stage to compensate for the beam path inside the NOPA. It is then focused by a plan-convex lens ($f = 300$ mm, 775 nm anti-reflex coating, Laser Components). The output of the NOPA is used without compression. It is focused by a concave aluminum mirror ($f = 400$ mm, focusing mirror 1 in Fig. 3.1). The focus is nearly collinearly overlapped with the focus of the 775 nm beam by a silver half-mirror. For the purpose of upconverting the NOPA output with the 775 nm beam in an SFM process (see Chap. 2.1.1), a BBO crystal (100 µm, cut at $\vartheta = 35°$, Döhrer) is placed into the nearly collinear beam path between 10 and 15 mm before the joint focus of both beams to avoid damaging and higher-order effects in the BBO. The 775 nm beam as well as the NOPA output are p-polarized (ordinary polarization). To achieve a better time-overlap between the pulses of 160 fs from the 775 nm beam path and the highly chirped, uncompressed NOPA output of potentially > 200 fs, the 775 nm beam is directed through 45 mm of SF10 (glass block in Fig. 3.1) and thereby chirped (see Chap. 2.1.1).

For initial alignment, the BBO is tilted (ϑ is tuned) such that both the second harmonic of the NOPA pulse and the second harmonic of the 775 nm pulse can be weakly observed. To facilitate identification of the reflex originating from the SFM process, an FS prism is placed in the beam path behind the BBO. It spatially separates the desired SFM reflex from reflexes of the fundamentals and second harmonics of the NOPA and the laser output. The SFM reflex is found by simultaneous fine-tuning of the manual delay stage and ϑ. Its energy is then optimized by tuning the delay stage, ϑ, focusing mirror 1, and the half mirror. Pulse energies between 0.8 and 2 μJ at wavelengths between 300 and 370 nm were achieved.

The light originating from the SFM process is collimated by a second aluminum focusing mirror ($f = 400$ mm, focusing mirror 2 in Fig. 3.1). Since the beam geometry at SFM is only nearly collinear, remaining fundamentals and second harmonics from the NOPA and the 775 nm beam are automatically spatially separated from the SFM beam path and can be blocked several tens of centimeters behind the BBO by an iris aperture. After collimation, the SFM pulses are directed into a prism compressor consisting of two FS prisms with an apex angle of 45° (Laser Components) and a retroreflecting aluminum mirror to compensate for GVD due to the NOPA process, the glass block, and the SFM process. Since the UV pulses exhibit s-polarization (extraordinary polarization, see Chap. 2.1.1), the prisms are provided with a broadband anti-reflex coating for s-polarized UV pulses. The UV pulses are coupled out of the compressor by an aluminum half mirror and directed to the sample. The length of the UV pulses was characterized by cross-correlation with probe pulses of known length in a suitable fluorescence dye solution to be < 95 fs (for details see Sect. 3.3).

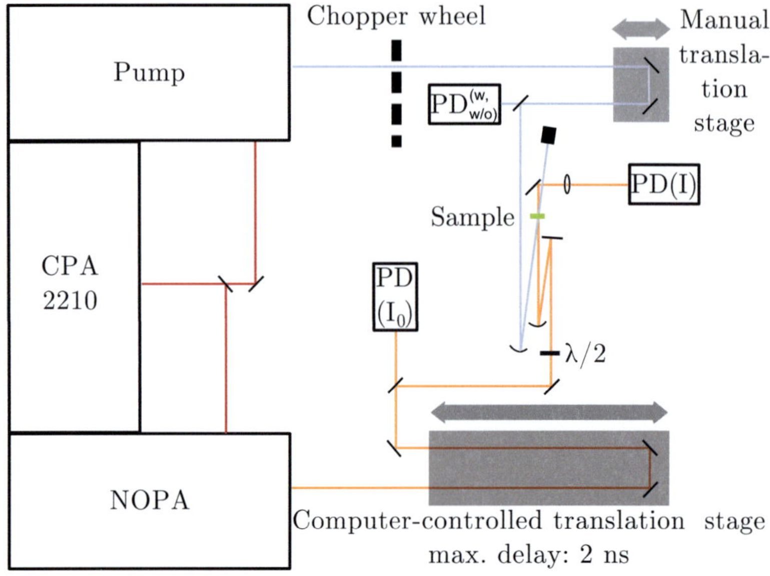

Figure 3.2.: Schematic picture of the employed experimental setup. Pump pulses (blue) were either generated by a UV-NOPA (see Sect. 3.2) or by SHG. Probe pulses (orange) were generated by a NOPA.

3.3. Experimental setup

The experimental setup is schematically depicted in Fig. 3.2. From the output of the CPA 2210, 250 µJ are used to pump a commercial NOPA system (Clark-MXR). The output of this NOPA, which is used for probing the sample, is directed over a computer controllable translation stage (Physik Instrumente). It allows for a maximum delay of the probe pulses by 2 ns (30 cm). Since it is unfavorable to align the translation stage in a way that the time overlap between pump and probe pulses is achieved only at minimum or maximum deflection, the maximum usable delay time is about 1.6 ns. The beam is then partly reflected by a broadband beam splitter (Laser Components). Its polarization plane is rotated to magic angle (54.7°) with respect to the pump pulse polarization by a wave plate

(Karl Lambrecht), and the beam is focused into the sample by a concave aluminum mirror ($f = 400$ mm). Behind the sample, the beam is recollimated and directed into a photodiode (Hamamatsu), labeled as PD(I) in Fig. 3.2. Together with a second photodiode behind the broadband beam splitter, labeled as PD(I_0), it allows for the measurement of the OD $\left(\lg \frac{I_0}{I} \right)$ of the sample.

Pump pulses are generated by the UV-NOPA setup described in Sect. 3.2. For pumping at $\lambda = 388$ nm, which cannot be achieved by the UV-NOPA, the second harmonic of the laser output is generated employing the same setup with the following minor modifications (see Fig. 3.1): The glass block is removed and the beam path to the NOPA blocked. Additionally, the BBO angle is tuned to allow for SHG at 775 nm. After recollimation of the transmitted fundamental and second harmonic, instead of compressing the pulses, they are sent directly to the sample by employing three dielectric mirrors (Laser Components) with high reflectivity for the second harmonic and high transmittivity for the fundamental. Thereby, the fundamental is efficiently removed from the beam path.

The repetition rate of the pump pulses is reduced to 500 Hz by a chopper wheel (EG & G). A small part of the pulse energy is directed into another photodiode (PD(w,w/o)) to detect the pulses which are transmitted by the chopper wheel. Thus, the relative optical density of the sample dependent on the pump-probe delay ($\Delta OD(\tau)$) can be evaluated according to

$$\Delta OD\left(\tau\right) = OD_w\left(\tau\right) - OD_{w/o} = \left(\log \frac{I_0}{I\left(\tau\right)} \right)_w - \left(\log \frac{I_0}{I} \right)_{w/o}, \qquad (3.2)$$

whereat $OD_w(\tau)$ and $OD_{w/o}$ are always obtained from consecutive probe pulses. The remaining pulse energy is sent over a manual translation stage for alignment purposes and focused by a concave aluminum mirror ($f = 500$ mm) into the sample, spatially overlapping with the probe pulses.

3.4. Experimental conditions

The spot sizes of pump and probe pulses were of 2 and 1 mm diameter, respectively. Thus, the sample volume irradiated by the probe pulse always fully overlapped with the volume irradiated by the pump pulse. The pump intensity was always kept below $1.7 \cdot 10^8$ W / cm^2 to ensure that only small changes in ΔOD occured. The temporal width of the probe pulses was measured by an intensity autocorrelator (Clark-MXR, for details see e.g. Ref. 38) to be ≈ 30 fs in the visible and ≈ 50 fs in the NIR spectral range. The overall time resolution of the experiment was evaluated by cross correlation of pump and probe pulses in a suitable fluorescence dye solution of the same OD at λ_p as the samples.

Fluorescence dyes exhibit excited state lifetimes on the order of ns and therefore can be expected not to display dynamics on the time scale of the experiment in their TA traces. Thus, they exhibit zero ΔOD at negative delay times, since in this case the probe pulse arrives at the sample before the pump pulse, and a time-independent ΔOD value A at positive delay times. The time resolution of the experiment is expressed by the rise time of ΔOD in the direct vicinity of $\tau = 0$ fs, which can be analyzed by an instrument response function $g(\tau)$. Assuming a Gaussian temporal shape of the fs pulses, the instrument response function has the form of an error function[81]

$$g(\tau) = \left(1 + \mathrm{erf}\left[\sqrt{4\ln(2)}\,\frac{\tau}{\tau_0} \right] \right). \tag{3.3}$$

The argument of the error function τ_0, which is an appropriate measure of the experimental time resolution, is connected to the FWHM of pump and probe pulses by

$$\tau_0 = \sqrt{\Delta t_p^{\,2} + \Delta t_{pr}^{\,2}}. \tag{3.4}$$

Since Δt_{pr} can be obtained independently, the length of the UV pump pulses can be evaluated from τ_0. Thus, the fit function for analysis of the cross-correlation measurements is

$$\Delta OD(\tau) = g(\tau)A. \tag{3.5}$$

All samples were measured as solutions at 19 °C in FS cuvettes (Hellma Analytics) with an optical path length of 1 mm. In some cases, where rapid degradation of the sample was observed, a flow cell system consisting of a 1 mm cuvette (Hellma Analytics) and a gear pump (Micropump) was employed. To circumvent reduction of the time resolution by GVM, sample concentrations were – if possible – chosen to exhibit an OD of 3 at λ_p. As can be calculated by Eq. 3.1, in a sample with an optical pathlength of 1 mm and an OD of 3, 75 % of the photons are absorbed within the first 200 µm of the optical path. Thus, the effective optical path can be assumed to be 200 µm. Comparison with literature-known values for GVM between λ_p = 319 nm and λ_{pr} = 589 nm of 65 fs (cyclohexane), 45 fs (methanol), 52 fs (ethanol), and 56 fs (isopropanol)[82] for an optical pathlength of 200 µm yields that the reduction of the time resolution by GVM is well below 100 fs.

3.5. Samples

The solvents cyclohexane, isopropanol, chloroform, ethanol, and methanol (spectral grade) were purchased from Carl Roth. Trichloroethanol ($\geq$ 99%) was purchased from Sigma Aldrich.

3.5.1. Halogenated cyclopentadienes

C_5Cl_6 (purity: 99 %) was obtained from Dr. Ehrenstorfer and used without further purification. C_5Br_6 was synthesized according to the procedure in appendix A.1, which is adapted from Ref. 83. A ^{13}C-NMR spectrum is depicted in Fig. A.1. The purity is estimated to be > 95 %.

Transient absorption (TA) spectroscopy

The employed λ_p were 323 nm for C_5Cl_6 and 350 nm for C_5Br_6. The range of λ_{pr} was between 500 nm and 1050 nm. The time resolution was measured by cross-correlation in the dyes 4,4'''-Bis-(2-butyloctyloxy)-p-quaterphenyl (BBQ) ($\lambda_p = 323$ nm) and 2,2'-([1,1'-Biphenyl]-4,4'-diyldi-2,1-ethenediyl)-bis-benzenesulfonic acid (stilbene 3) ($\lambda_p = 350$ nm) to be always below 100 fs. The maximum delay between pump and probe pulses was 1.6 ns. Samples were prepared as solutions in cyclohexane, isopropanol, chloroform, and trichloroethanol and measured in the flow cell system, because rapid photodegradation of the samples was observed. To reduce effects of group- velocity mismatch (GVM), the OD of the solutions was kept above 3 at the respective excitation wavelengths. The concentrations were thus in the range of $2.5 \cdot 10^{-2}$ mol/l.

Time-resolved photoelectron (TRPE) spectroscopy

TRPE spectroscopy was conducted in collaboration with O. Schalk, G. Wu and P. Lang in the group of A. Stolow at the National Research Council of Canada, Ottawa. The employed TRPE spectrometer consists of a femtosecond laser system and a magnetic bottle time of flight spectrometer in combination with a supersonic molecular beam. The experimental setup is described in detail in Ref. 84. Pump pulses at $\lambda_p = 315$ nm were generated by frequency quadrupling of the output of an optical parametric amplifier. Pulse energies were 1.5 μJ (at $\lambda_{pr} = 267$ nm) and 1.6 μJ (at $\lambda_{pr} = 400$ nm). For the probe wavelengths $\lambda_{pr} = 267$ nm (3.2μJ) and 400 nm (16 μJ) a part of the laser output at 800 nm was frequency tripled or frequency doubled, respectively. The relative polarization of pump and probe pulses was rotated to magic angle by a Berek compensator. The pulses were collinearly focused into the interaction chamber of the photoelectron spectrometer. Since the sample was to be ionized by 2- or 3-photon absorption, the light intensities in the interaction regions were by orders of magnitude higher than in the TA experiment. The temporal cross correlation of pump and probe pulses was measured in NO to be 140 fs. In the interaction chamber a supersonic molecular beam generated by a 1 kHz Even-Lavie valve with a 250 µm diameter conical nozzle perpendicularly crossed the optical beam path. Helium was used as a carrier gas with a backing pressure of 3.8 bar. C_5Cl_6 was introduced into the body of the valve as a liquid, C_5Br_6 as a solid. C_5Cl_6 was slightly heated to 40 °C, C_5Br_6 to 80 °C, to enhance the vapor pressure. To prevent condensation of the sample in the nozzle of the valve, the front plate of the valve was separately heated to the same temperature. Photoelectron energies were calibrated by using the known photoelectron spectrum of NO.[85] The maximum pump-probe delay was 300 ps.

3.5.2. Type I Photoinitiators

Benzoin (Bz) was purchased from Sigma-Aldrich, 2,4,6-trimethylbenzoin (TMB) and mesitil (Me) were synthesized by D. Voll at the Institute for Chemical Technology and Polymer Chemistry, Karlsruhe Institute of Technology, according to literature known procedures.[7,86–90]

Transient absorption (TA) spectroscopy

The employed λ_p was chosen to be the same as in earlier PLP experiments,[7] i.e. 351 nm. Three different λ_{pr} of 470 nm, 500 nm and 600 nm were employed, whereas mesitil did not show any TA at λ_{pr} = 600 nm. For allowing comparison of the samples it was mandatory to prepare them with the same OD. However, the extinction coefficients of the three samples at λ_p differ, at the most, by a factor of 100 (see Fig. 5.3). The concentrations at an OD of 3 therefore would have been in a range, where bimolecular effects like clustering and bimolecular quenching have to be taken into account. Thus, a middle ground had to be found between concentration and time resolution. Therefore, the OD of the samples was chosen to be 0.5 at

$\lambda_p = 351$ nm. The resulting concentrations were therefore $7.8 \cdot 10^{-2}$ mol/l (Bz), 1.9 mol/l (TMB), and $4.9 \cdot 10^{-3}$ mol/l (Me). The reduced OD of the samples lead to a reduction of the experimental time resolution to 200 fs. Since no photodegradation of the samples could be observed in the absorption spectra, the experiments were performed in static cuvettes with an optical path length of 1 mm.

3.5.3. Photodepletable triplet photoinitiators

DETC and ITX were purchased from Exciton and Aldrich, respectively. Fluorescence spectra were measured by J. Fischer. The detailed experimental conditions are included in Ref. 91.

For TA spectroscopy, samples were prepared as solutions in ethanol with an OD of 0.3 at the excitation wavelength of the experiments. The concentrations of DETC and ITX were therefore $2.22 \cdot 10^{-4}$ mol/l and $3.8 \cdot 10^{-3}$ mol/l, respectively.

Transient absorption (TA) spectroscopy

In earlier lithography experiments of Fischer *et al.*,[92] the photoinitiators had been excited by 2 photons of 810 nm. A 2 photon excitation was not possible during the TA experiments due to different focusing and properties of the laser pulses. Instead the samples were excited by a 1 photon process at $\lambda_p = 388$ nm, the second harmonic of the laser output. The range of λ_{pr} was $473 - 900$ nm. The effect of GVM rapidly decreases with red-shift of λ_p to 388 nm (see e.g. the values listed in Ref. 82). Additionally, the time resolution was limited by the choice of the second harmonic of the 160 fs laser output as pump pulses. Therefore, the OD of 0.3, which was chosen for the sample solutions, did not have an effect on the time resolution. As preliminary photolysis experiments showed a high photostability of both samples in absence of any monomer, and TA test experiments with a flow cell showed no differences to the data taken from a cuvette, the experiments were performed in 1 mm fused silica cuvettes.

3.6. Theoretical methods

All DFT, TDDFT, HF, CC2, and CCSD calculations were performed employing the TURBOMOLE V-6.3 program package.[93] Geometry optimizations in the lowest singlet and triplet states were performed with DFT/BP86[94–97] and the def2-SV(P) basis set.[98–100] The combination is known to yield reliable minimum geometries.[101] CC2 calculations[102–104] were performed employing the aug-cc-pVDZ basis set[105,106] to obtain energies and first-order properties like dipole moments. The basis set proved to be a good compromise between accuracy and computational demand. It was used with augmentation by diffuse functions to be also able to treat Rydberg states with reasonable accuracy.[53] For DFT energy calculations and TDDFT calculations[107–110] the B3LYP functional in combination with the same basis was employed. B3LYP is known to give accurate results for excitation energies in a great variety of systems.[111,112] Ionization potentials (IPs) were calculated either with CCSD (C_5Cl_6) or, where the computational demand of CCSD was disproportional (C_5Br_6), with DFT/B3LYP. In both cases the def2-TZVP basis[99,113,114] was employed.

SA-CASSCF calculations were performed with the COLUMBUS V5.9 program package[115–118] employing a 6-31G* basis set.[119,120]

4. Photoinduced dynamics of halogenated cyclopentadienes

4.1. Introduction

The systems, which are investigated in this chapter, exhibit the shortest excited state lifetime of all systems within the present work. Their excited states dynamics can be discussed with the help of a Jabłoński-type diagram adapted from Fig. 1.1, which is shown in Fig. 4.1. The only process following excitation, which is relevant in the present chapter, is highlighted. The electronic state, which is excited by the pump pulse of the time-resolved experiment, is depopulated too rapidly to allow the molecules to fluoresce or to undergo ISC. The dynamics are, therefore, governed by IC into the electronic state of a reaction product.

Halogenated hydrocarbons are common environmental pollutants. In nature they are not only observed in the gas phase but also in solution dissolved in rivers and lakes or in cloud droplets, where they naturally are irradiated by sunlight.[121] Absorption of light can lead to degradation via formation of halogen and hydrocarbon radicals. Especially chlorine radicals are very reactive species, which exhibit a lifetime of several ps in solution[122, 123] and are potentially dangerous to biologic structures.[124] Halogen radicals tend to form charge-transfer (CT) complexes with solvent molecules.[123, 125–133] These complexes exhibit absorption bands in the visible and the UV. However, until now a microscopic picture of the radical-solvent interaction has not been fully established. Especially the stability of individual CT complexes between radicals and solvent molecules has not been investigated so far. Therefore, it is unknown if the CT complexes are

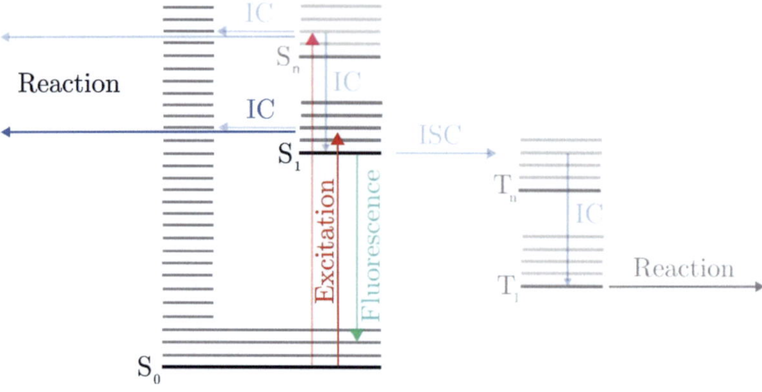

Figure 4.1.: Jabłoński-type diagram adapted from Fig. 1.1. The only excited state process relevant for the present chapter, direct reaction via ultrafast radiationless IC, is highlighted.

of a transient nature similar to solvent interactions like hydrogen bonds or, if a CT complex represents a more stable interaction between the radical and one or more specific solvent molecules.

For an approach to this problem the experimental setup has to be chosen such that the chemical system, which is irradiated, provides the possibility for the produced halogen radicals to form a geminate CT complex with the other product of the photoinduced bond fission reaction. Geminate recombination of the radical species is thereby an unwanted side reaction and has to be prevented. Systems which meet these demands are fully chlorinated (C_5Cl_6) and brominated (C_5Br_6) cyclopentadienes. It was demonstrated earlier that they undergo photoinduced homolytic dissociation of a carbon – halogen bond at the sp^3 position of the five-membered ring in solution to form a cyclopentadienyl and a halogen radical (see Fig. 4.2).[134–136] Furthermore, when irradiating frozen solutions of C_5Br_6 at 77 K, besides absorption bands of the C_5Br_5 radical an additional broad and intensive absorption band peaking at 480 nm was observed. This absorption band disappeared upon melting the frozen solvent and therefore could be assigned

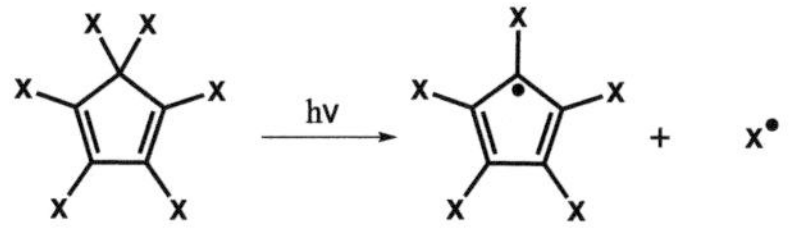

Figure 4.2.: Photoinduced reaction observed earlier for C_5Cl_6 and C_5Br_6 in solution.[134, 135]

to a geminate CT absorption band of the two radicals, which were trapped in the frozen solvent cage.

Secondary, the influence of the solvent on the properties of the complex can be investigated. Especially the interaction between the chlorine radical and chlorinated solvents is a potentially interesting issue. For this purpose a set of solvents has to be chosen, which cover a wide range of viscosity, polarity, and density values (see Tab. 4.1). Furthermore, suitable solvents should on the one hand not exhibit absorption in the near UV and on the other hand not form CT absorption bands with the halogen radicals overlapping with the geminate CT absorption band. Cyclohexane, isopropanol, chloroform, and trichloroethanol are employed, which proved to be a good choice in earlier experiments with cycloheptatriene.[43, 144] Cyclohexane exhibits zero polarity, low viscosity, and low density,[137] and does not show absorption in the investigated spectral range.[145] It is known not to exhibit CT absorption bands with halogen radicals in the visible spectral regime.[122, 128, 130] Isopropanol has a substantially higher polarity and

solvent	density / $g \cdot cm^{-3}$	dipole moment D	viscosity / mPas
cyclohexane	0.78^{137}	0.00	0.98^{137}
isopropanol	0.78^{138}	1.61	2.5^{138}
chloroform	1.47^{139}	1.04^{140}	0.56^{139}
trichloroethanol	1.56^{141}	2.04^{142}	21^{143}

Table 4.1.: Properties of the employed solvents at 20 °C. If not labeled different, dipole moments are calculated by CC2.

viscosity than cyclohexane.[138] It shows only very low absorption in the UV starting at 250 nm[146] and no CT absorption bands with Br radicals in the visible spectral range.[129] Chloroform is characterized by a low viscosity, high density, and low polarity.[139,140] It exhibits absorption only below 220 nm in the UV.[147] CT absorption bands with chlorine radicals are only observed in the UV.[127] Trichloroethanol exhibits high viscosity and polarity.[141–143] Its spectrum shows only moderate absorption below 250 nm. No CT absorption bands with halogen radicals have been reported to date for trichloroethanol.

Cyclopentadiene (C_5H_6) can be regarded as a cis-butadiene type polyene model system, since only its σ-backbone is cyclic, not its π-system. Accordingly, it shows a rich chemistry of pericyclic reactions in the framework of the Woodward-Hoffmann rules.[148–151] At room temperature it spontaneously dimerizes via a Diels-Alder reaction. According to the Woodward-Hoffmann rules a electrocyclic reaction to form bicyclo[2,1,0]pentene is thermally allowed in a conrotatory, and photochemically allowed in a disrotatory fashion. Additionally, a photochemically allowed [1,3]-sigmatropic and a thermally allowed [1,5]-sigmatropic suprafacial hydrogen shift is predicted. Bicyclo[2,1,0]pentene and an additional species, tricyclo[2,1,0,0]-pentane were experimentally observed after irradiation of C_5H_6 in solution.[149–151] The observation of the sigmatropic reactions is extremely difficult, since the reaction products differ from the educts only by a pseudorotation. Moreover, photoinduced H atom abstraction was observed.[152]

As opposed to C_5Cl_6 and C_5Br_6 excited states and exited states dynamics of C_5H_6 and its methylated derivatives have been subject to many experimental[153–162] and theoretical[163–171] investigations. It is known for long to undergo ultrafast radiationless relaxation after photoexcitation in the gas phase.[155,159,161] In many polyenes ultrafast relaxation dynamics take place via a spectroscopically dark state with (partly) doubly excited character, which connects the excited bright state to the ground state by conical intersections.[56,172–180] In C_5H_6 the role of a low lying, spectroscopically dark and multiconfigurational excited singlet state with considerably doubly excited character is, however, a subject of current discussion.[171] Hints for the existence of a [1,3]-sigmatropic hydrogen shift reaction could be found by comparing the dynamics of C_5H_6 with its deuterated analogue in the gas phase.[159] More recent quantum chemical dynamics simulations in the excited states though, point to the fact that other reaction channels are more favored.[171] In fact, the optimized minimum energy conical intersection geometries obtained in this study are in agreement with a disrotatory "transition state" of an electrocyclic reaction. However, sigmatropic shift reactions have to be taken into account, especially in the investigation of the photochemistry of halogenated derivatives, since sigmatropic hydrogen as well as chlorine shifts were observed in a similar system, cycloheptatriene.[43, 144, 181]

Only the spectrum of C_5Cl_6 is literature-known.[182] The dynamics of the photoinduced bond dissociation reaction of C_5Cl_6 and C_5Br_6 have not been investigated. Since their π-electronic structure is in analogy to C_5H_6, the possibility of additional reaction channels as observed for C_5H_6 is high. Thus, to be able to compete with such ultrafast channels, the bond dissociation channel observed for C_5Cl_6 and C_5Br_6 has to take place on a similar time scale. Hence, the bond dissociation, the subsequent formation of CT absorption bands and their fate are to be investigated by femtosecond transient absorption (TA) and time-resolved photoelectron (TRPE) spectroscopy supported by quantum chemical calculations.

4.2. Excited states at the Franck-Condon geometry

In analogy to C_5H_6, the halogenated cyclopentadienes C_5Cl_6 and C_5Br_6 exhibit a ground state minimum (Franck-Condon) geometry with C_{2v} symmetry. The Cartesian coordinates of the optimized structures are found in Tabs. C.1 and C.4 in the appendix C. The absorption spectrum of C_5Cl_6 in cyclohexane is depicted in Fig. 4.3 (a). The first absorption band of C_5Cl_6

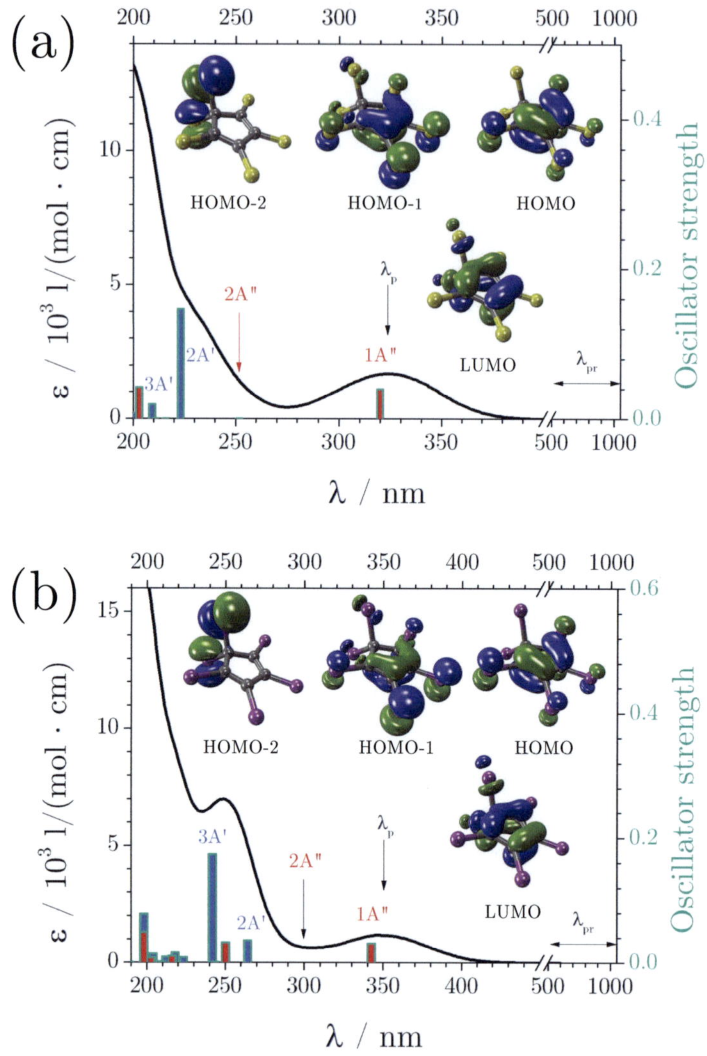

Figure 4.3.: Absorption spectra of (a) C_5Cl_6 and (b) C_5Br_6 in cyclohexane solution. For comparison transitions to states with A' (blue) and A" symmetry (red) calculated with CC2 are shown. Transitions to selected excited states are additionally labeled. Thereby, transitions with weak oscillator strength are marked with arrows. Inserted are visualizations of the three highest occupied MOs and the LUMO and, additionally, the λ_p and the λ_{pr} range.

is centered at 323 nm. Since the first absorption band of C_5H_6 is centered at 235 nm,[156] the chlorine substitution obviously leads to a substantial red-shift of the spectrum. In addition to the experimental spectrum of C_5Cl_6, singlet excitations calculated with CC2 at the Franck-Condon geometry are inserted as bars in Fig. 4.3 (a) (for details on the calculation see Chap. 3.6). For the purpose of describing both the Franck-Condon geometry and the minimum geometries of the photoproducts of C_5Cl_6 within the same symmetry group, the nomenclature of the electronic states is not according to the C_{2v}, but to the C_s symmetry group. The overall agreement between calculated and experimental spectra is very good. The discussion of the calculated excited states will be in the following reduced to the ones which are explicitly labeled in Fig. 4.3 (a). The first absorption band of C_5Cl_6 can be exclusively assigned to a transition from the closed shell ground state (1A') to the lowest state with A" symmetry (1A"). The 1A" state is – like in C_5H_6 – governed by a single-electron $\pi\pi^*$ excitation from the highest occupied MO (HOMO) to the lowest unoccupied MO (LUMO) (see also the MOs depicted in Fig. 4.3 (a)). The second lowest state with A" symmetry (2A") is calculated to exhibit negligible oscillator strength and to posses strong Rydberg character, i.e. it corresponds to a single-electron transition from the HOMO to a highly diffuse Rydberg MO.[183] The second lowest state of C_5Cl_6 with A' symmetry (2A') is calculated to be governed by a LUMO←HOMO-1 transition. This state probably corresponds to the second excited valence state in C_5H_6 mentioned in Sect. 4.1, which is known to exhibit strong multiconfigurational character with considerable contribution from a doubly excited configuration.[166,171] Since CC2 poorly treats such states[184] (see Chap. 2.2), its energy may be substantially overestimated. Earlier experimental studies proposed a participation of the 2A' state in the excited states dynamics of C_5H_6,[159,161,162] which would make a more precise description of the corresponding state in C_5Cl_6 desirable. However, a recent theoretical study partially relativizes the role of this state.[171] Furthermore, as will be shown below, the state does not play

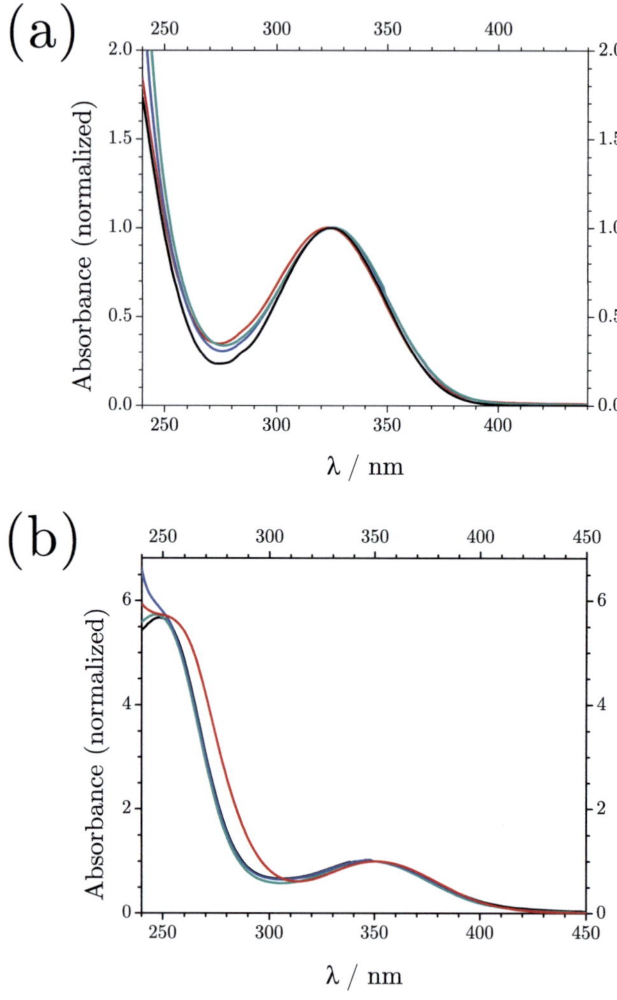

Figure 4.4.: Absorption spectra of (a) C_5Cl_6 and (b) C_5Br_6 in cyclohexane (black), isopropanol (green), chloroform (blue), and trichloroethanol (red). The spectra are normalized with respect to the absorbance at (a) 323 nm and (b) 350 nm.

a significant role in the excited states dynamics of C_5Cl_6. The 3A' state is governed by a single electron LUMO$\leftarrow$HOMO-2 excitation. Since the HOMO-2 (see Fig. 4.3 (a)) can be described as a linear combination of two p atomic orbitals (AO) of the chlorine substituents at the sp^3 carbon, this state has no analogue in C_5H_6. Additionally, as the earlier observed photoinduced bond dissociation takes place between the sp^3 carbon and one of the chlorine substituents (see Fig. 4.2), the energy of the 3A' state is expected to be highly influenced by the bond dissociation.

The absorption spectrum of C_5Br_6 in cyclohexane is shown in Fig. 4.3 (b). Its first absorption band at 350 nm is further red-shifted with respect to C_5H_6 than in the case of C_5Cl_6. As can be seen from the MO visualizations in Fig. 4.3 (b), the electronic structure is comparable to C_5Cl_6. Also the states, which are labeled in Fig. 4.3 (b), have the same character. Only the 2A' and 3A' states interchange their characters. The LUMO$\leftarrow$HOMO-2 transition is the 2A' state in C_5Br_6 and the LUMO$\leftarrow$HOMO-1 transition the 3A' state.

In contrast to the influence of halogen substitution, the solvent influence on the absorption spectra (see Fig. 4.4) is rather small. Since C_5Cl_6 and C_5Br_6 exhibit very low dipole moments (0.69 and 0.53 D, calculated with CC2) in the ground state, it is not expected to be substantially influenced by solvation. However, the respective 1A" states exhibit by far higher calculated dipole moments (2.38 and 2.91 D) and therefore a higher effect of solvent polarity on the spectra especially between the solvents cyclohexane (0 D) and trichloroethanol (2.04 D[142]) would be expected (for additional solvent properties see Tab. 4.1).

4.3. Transient absorption (TA) spectroscopy in solution

The transient response of the two molecules C_5Cl_6 and C_5Br_6 to excitation of the 1A" state in solution is now investigated by TA spectroscopy employing a wide range of λ_{pr} from 500 nm to 1050 nm.

C_5Br_6

The features, which can be observed in the TA traces of C_5Br_6, are only slightly solvent dependent. They are therefore discussed with the help of the data taken in cyclohexane, which are shown in Fig. 4.5 (a). The solvent-dependent differences are discussed afterwards.

The population initially prepared in 2A" at time zero by the pump pulse yields positive ΔOD values irrespective of the employed λ_{pr} due to ESA. ΔOD then decays within < 100 fs. After the initial decay the evolution of TA is highly dependent on λ_{pr}: At short λ_{pr} (500 nm - 600 nm in cyclohexane) TA grows again on a time scale of a few tens of ps and decays on a time scale, which is beyond the time window investigated by the experiments (1.6 ns). At long λ_{pr} (650 nm - 900 nm in cyclohexane) the TA only decays on a time scale, which is comparable to the rise at short λ_{pr}. Some remaining TA intensity also decays on a time scale beyond 1.6 ns.

In the other solvents isopropanol, chloroform and trichloroethanol (see Figs. 4.5 (b) and 4.6) a similar behavior is observed. In isopropanol the growth of TA on a ps time scale is observed in the same range of λ_{pr}, but the time scale of the growth exhibits a slight shift to longer delay times at longer λ_{pr}. This is also observable by the slight shift of the maxima on a ps time scale, which is marked by arrows in Fig. 4.5 (b). In chloroform such a shift in the TA maxima cannot be observed. In comparison to the solvents cyclohexane and isopropanol the range of λ_{pr}, where the growth of TA is observable, is enlarged to 500 - 700 nm. The same spectral broadening of TA growth as in chloroform can be also observed in trichloroethanol. Moreover, a λ_{pr}-dependent shift of the TA maxima (see the arrows in Fig. 4.6 (b))

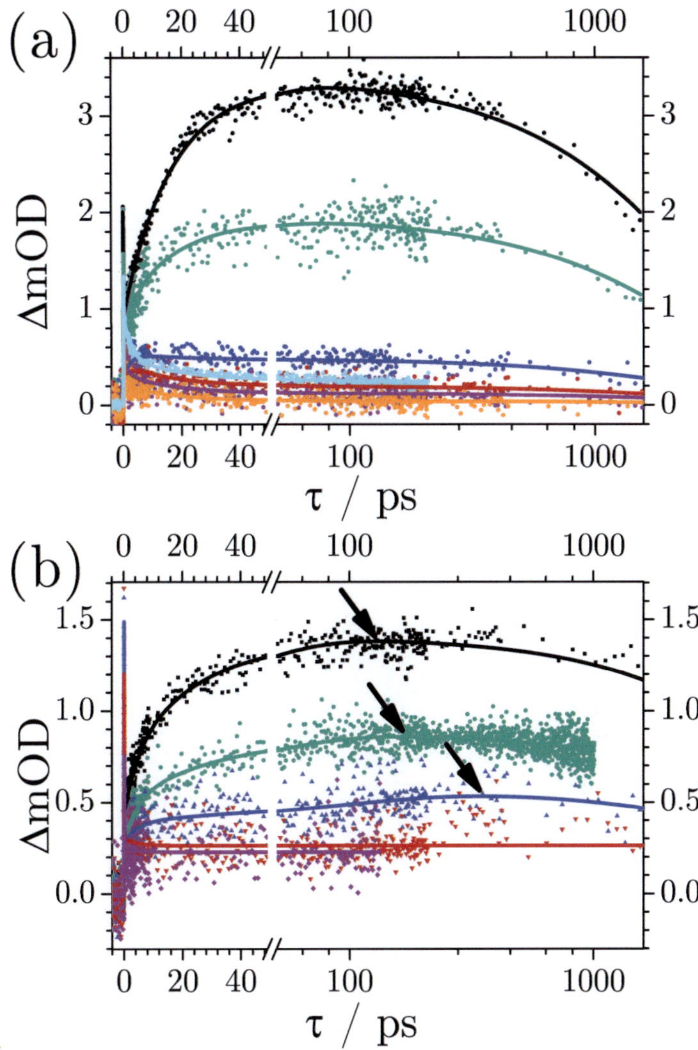

Figure 4.5.: TA traces (symbols) of C_5Br_6 in (a) cyclohexane and (b) isopropanol together with global fit analysis (lines). All data refer to $\lambda_p = 350$ nm. The probe wavelengths are $\lambda_{pr} = 500$ nm (black), 550 nm (green), 600 nm (blue), 650 nm (red), 700 nm (violet), 750 nm (orange), and 900 nm (light blue). Note the shift of the TA maximum to longer delay times at longer λ_{pr} in the case of the sample in isopropanol as marked by arrows.

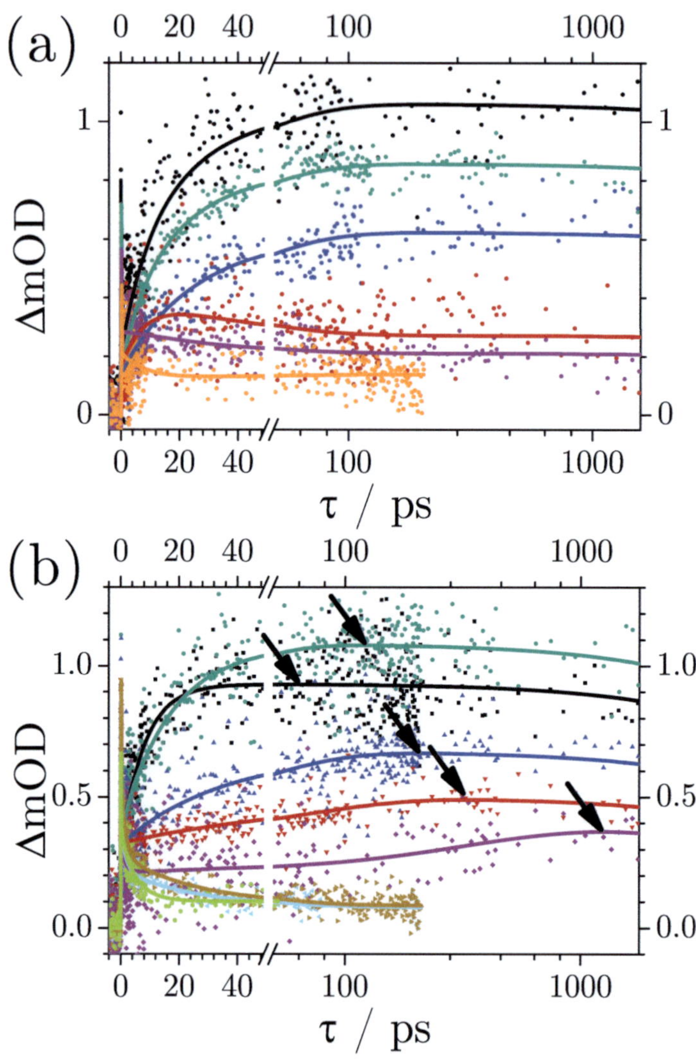

Figure 4.6.: TA traces (symbols) of C_5Br_6 in (a) chloroform and (b) trichloroethanol together with global fit analysis (lines). All data refer to $\lambda_p = 350$ nm. The probe wavelengths are $\lambda_{pr} = 500$ nm (black), 550 nm (green), 600 nm (blue), 650 nm (red), 700 nm (violet), 750 nm (orange), 900 nm (light blue), 950 nm (dark yellow), and 1000 nm (light green). Note the shift of the TA maximum to longer delay times at longer λ_{pr} in the case of the sample in trichloroethanol as marked by arrows.

like in isopropanol is observable. In summary, the growth of TA seems to be spectrally broadened in chlorinated solvents, whereas a spectral shift of its maximum in time is only observed in the more viscous and polar solvents isopropanol and trichloroethanol (see Tab. 4.1).

For a quantitative picture of the dynamics observed in the TA traces, the data have to be analyzed by fit functions. Since the time scale of the observed dynamics is mostly in the range of ps, a statistical nature can be assumed for the observable processes. Moreover, the time scale points to predominantly unimolecular processes only influenced by the solvent cage. As will be discussed below, diffusion is too slow to play a significant role. Hence, the mathematical expression for unimolecular kinetics, an exponential decay function, is a meaningful choice for the evaluation of the data. Processes on the order of fs and single ps are, however, in general not statistical. Nevertheless, it is convenient also to fit such processes by exponential functions.[161, 180] Obviously, the observed dynamics are not attributable to a single process. Therefore, a sum of exponentials is employed. Tests showed good results with a sum of 4 exponentials. With only

solvent	τ_1 / ps decay	τ_2 / ps decay	τ_3 range / ps rise	τ_4 / ns decay
cyclohexane	< 0.1	2.3 ± 0.3	17.3 ± 0.4	2.9 ± 0.1
isopropanol	< 0.1	5 ± 1	26 ± 1 – 110 ± 30	> 3.2
chloroform	< 0.1	9.0 ± 0.1	28 ± 3	> 3.2
trichloroethanol	< 0.1	1.6 ± 0.3	8.0 ± 0.8 – 400 ± 100	> 3.2

Table 4.2.: Time constants resulting from global fits of the C_5Br_6 TA traces in different solvents with a sum of 4 exponential functions. Additionally, it is mentioned, whether the time constants are associated to a rise or decay of the TA. Time constants are optimized globally. Only in the case of isopropanol and trichloroethanol is τ_3 optimized separately for each λ_{pr}.

3 exponentials the fits failed to at least qualitatively describe the data. With 5 exponentials the fits were in some cases unstable and did only give trivial improvements against 4 exponentials on the χ^2 values of the employed Levenberg-Marquardt fitting routine.

Thus, the data were fitted globally with functions of the type

$$\Delta OD\left(\lambda_{pr}, \tau\right) = g\left(\tau\right) \otimes \left[\sum_{i=1}^{4} A_i\left(\lambda_{pr}\right) \cdot P_i\left(\tau\right)\right], \qquad (4.1)$$

where $g(\tau)$ is the instrument response function according to Eq. 3.3, $P_i\left(\tau\right) = e^{-\frac{\tau}{\tau_i}}$ is an exponential decay function depending on a time constant τ_i and modeling a specific population evolution process. $A_i\left(\lambda_{pr}\right)$ is the decay associated difference spectrum connected to the process P_i and dependent on λ_{pr}. The 4 exponential functions can be associated to observable features in the TA traces: The initial decay of TA after time zero (τ_1), additional dynamics on the time scale of few ps (τ_2), the growth of TA on the time scale of few tens of ps (τ_3), and the ultimate decay of TA beyond the experimental time window (τ_4).

In the global fitting procedure the A_i are optimized separately for each λ_{pr}, the time constants τ_i globally for all λ_{pr}. Only in the solvents isopropanol and trichloroethanol, where the spectral shift of TA growth was observable, was τ_3 optimized separately for each λ_{pr}. The time constants resulting from the fits are listed in Tab. 4.2. Their errors are the standard deviations from the fitting routine or – if larger – the experimental time resolution. If the values are beyond the time resolution of the experiment, the time resolution is given instead as an upper boundary. Although the experimentally accessible time window is 1.6 ns, time constant values in the range of few ns seem to be in good agreement with published experimental results (see Chap. 6.3). Thus, time constants with values lower than the doubled time window (3.2 ns) are listed as obtained from the fits. For time constants with higher values only a lower boundary of 3.2 ns is given.

C_5Cl_6

As opposed to C_5Br_6, the TA traces of C_5Cl_6 in cyclohexane, isopropanol, chloroform, and trichloroethanol (see Figs. 4.7 and 4.8) show a, by far, more pronounced solvent dependence. In isopropanol and chloroform (Figs. 4.7 (b) and 4.8 (a)) the TA exhibits a time evolution qualitatively similar to the observations in C_5Br_6. After an initial decay of TA within < 100 fs at short λ_{pr} the TA grows again within a few tens of ps and eventually decays on a time scale at the edge or beyond the experimental time window. At long λ_{pr} only a decay of TA can be observed. Furthermore, the spectral range in which an intermediate TA growth is observable shows the same solvent dependence as in C_5Br_6. Only the time scales of growth and depletion of TA at short λ_{pr} differ. Moreover, as opposed to C_5Br_6, a λ_{pr}-dependent shift of the TA maximum in time is not observed in isopropanol.

In contrast, in the solvent trichloroethanol (Fig. 4.8 (b)) the shift is much more pronounced than observed in C_5Br_6 with the same solvent. Furthermore, as opposed to any other solute-solvent combination, the ultimate decay of TA takes place on a ps time scale.

solvent	τ_1 / ps decay	τ_2 / ps decay	τ_3 range / ps rise	τ_4 / ns decay
cyclohexane	< 0.1	1.1 ± 0.1		> 3.2
isopropanol	< 0.1	2.9 ± 0.5	12 ± 2	0.87 ± 0.04
chloroform	< 0.1	2.1 ± 0.6	15 ± 2	2.7 ± 2
trichloroethanol	< 0.1	1.7 ± 0.1	12 ± 1 – 140 ± 30	0.220 ± 0.002

Table 4.3.: Time constants resulting from global fits of the C_5Cl_6 TA traces in different solvents with a sum of 4 exponential functions. Additionally, it is mentioned, whether the time constants are associated to a rise or decay of the TA. Time constants are optimized globally. Only in the case trichloroethanol is τ_3 optimized separately for each λ_{pr}.

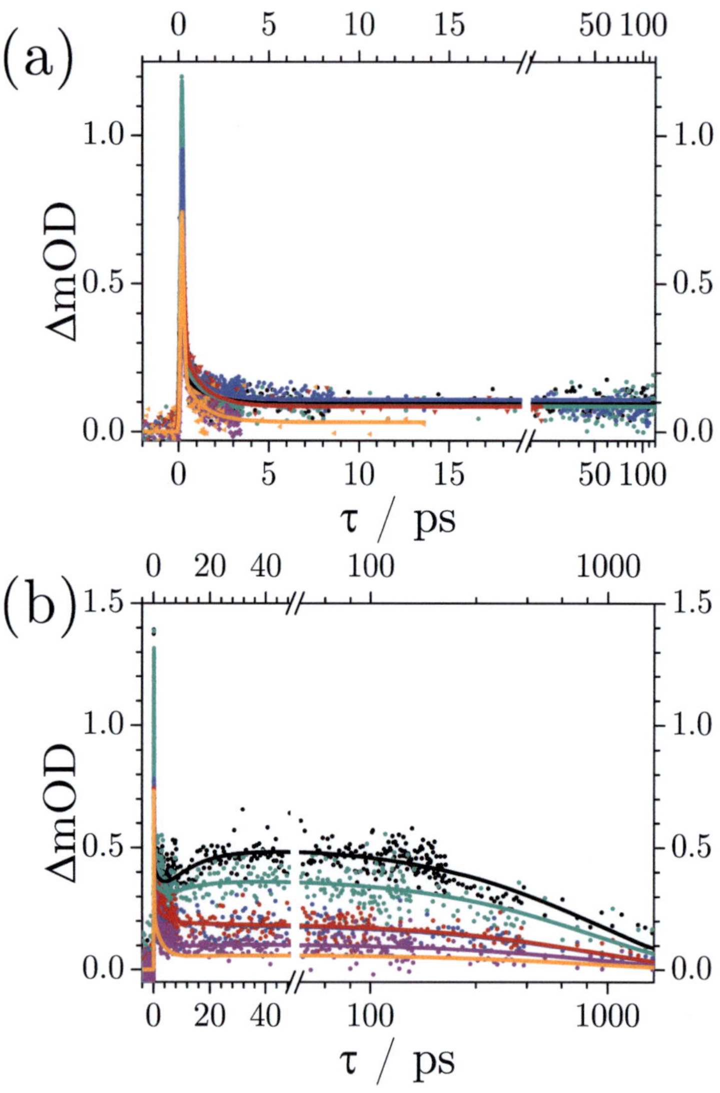

Figure 4.7.: TA traces (symbols) of C_5Cl_6 in (a) cyclohexane and (b) isopropanol together with global fit analysis (lines). All data refer to $\lambda_p = 323$ nm. The probe wavelengths are $\lambda_{pr} = 500$ nm (black), 550 nm (green), 600 nm (blue), 650 nm (red), 700 nm (violet), and 750 nm (orange).

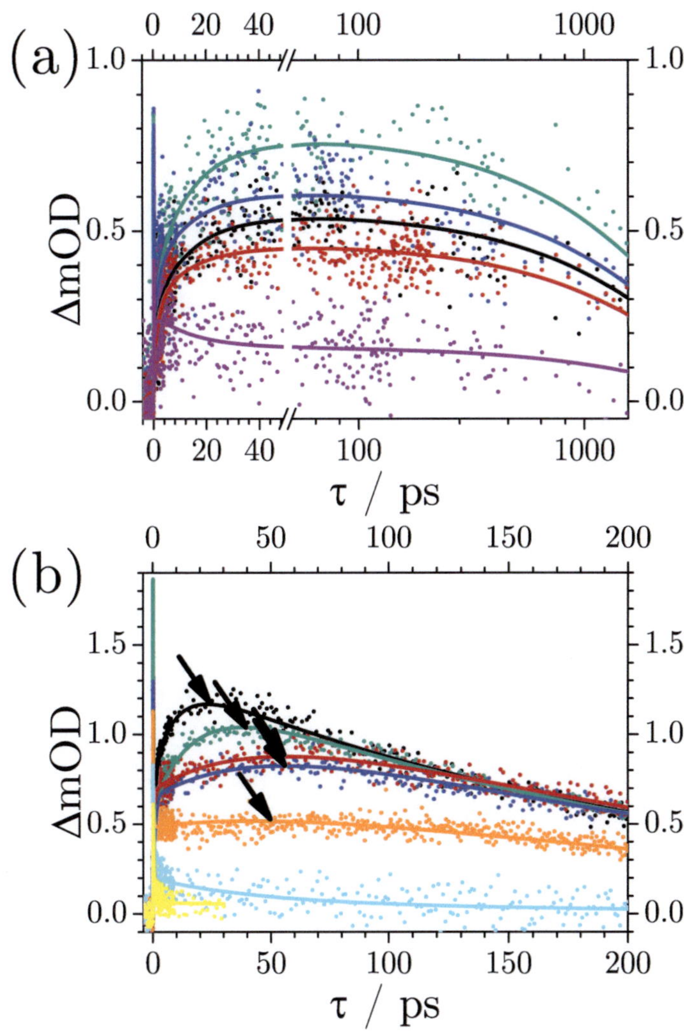

Figure 4.8.: TA traces (symbols) of C_5Cl_6 in together with global fit analysis (lines). All data refer to $\lambda_p = 323$ nm. In the solvent chloroform (a) : The probe wavelengths are $\lambda_{pr} = 500$ nm (black), 550 nm (green), 600 nm (blue), 650 nm (red) and 700 nm (violet). In the solvent trichloroethanol (b): $\lambda_{pr} = 500$ nm (black), 550 nm (green), 615 nm (blue), 650 nm (red), 720 nm (orange), 900 nm (light blue), and 1050 nm (yellow). Note the shift of the TA maximum to longer delay times at longer λ_{pr} as marked by arrows and the different time scale.

The TA traces in cyclohexane (Fig. 4.7 (a)) show also a qualitatively different behavior than in C_5Br_6. The initial TA intensity decays within < 100 fs like in the other solvents. In sharp contrast to the other samples no rise of TA intensity at short λ_{pr} on a ps time scale is observed.

The TA data were analyzed via the same fit functions (Eq 4.1) as employed for C_5Br_6. However, in the case of C_5Cl_6 in cyclohexane only 3 exponentials were needed to fit the data due to the missing growth of TA at short λ_{pr}. The resulting time constants are listed in Tab. 4.3.

4.4. Time-resolved photoelectron spectroscopy in the gas phase

To be able to distinguish purely excited states dynamics from solvent dependent dynamics in solution, C_5Cl_6 and C_5Br_6 were investigated by TRPE spectroscopy in the gas phase, which was conducted in collaboration with O. Schalk, G. Wu and P. Lang in the group of A. Stolow at the National Research Council of Canada, Ottawa.

In the gas phase excited molecules cannot dissipate the photoenergy to the environment due to ultra high vacuum conditions and very low translational temperatures.[185, 186] Upon return to the ground state, the absorbed energy is redistributed into vibrational degrees of freedom. Since C_5Cl_6 and C_5Br_6 exhibit a very limited number of these (27), they are highly vibrationally excited. Assuming a canonical distribution, the energy of one photon of $\lambda_p = 315$ nm is consistent with a characteristical vibrational temperature of 2066 K (C_5Cl_6) and 2019 K (C_5Br_6).[187] Such a high vibrational temperature leads to a poor Franck-Condon overlap between ground state and ionic states and therefore to negligible ionization cross-sections. Thus, photoelectrons from ionization of vibrationally "hot" molecules in the ground state are not expected to be observable. All observable features in the TRPE spectra, therefore, can be assigned to excited states or to reaction products.

For TRPE spectroscopy of C_5Cl_6 two λ_{pr} are employed, 267 and 400 nm. The TRPE spectra are shown in Fig. 4.9 (a) and (b). The IP of C_5Cl_6 is calculated by CCSD to be 8.89 eV. The value is in good agreement with the experimental IP of C_5H_6 of 8.57 eV.[188] Thus, at time zero, when pump and probe pulses temporally overlap, C_5Cl_6 can at $\lambda_{pr} = 267$ nm only be ionized by absorption of one pump photon (315 nm, 3.94 eV) and two probe photons (267 nm, 4.64 eV, [1,2'] absorption) and at $\lambda_{pr} = 400$ nm by absorption of one pump photon and 2 probe photons (400 nm, 3.10 eV). The maximum translational excess energy of the photoelectrons resulting from the ionization is thus the sum of the photon energies minus the IP, i.e. 4.33 eV and 1.25 eV, depending on λ_{pr}. Delaying the probe pulse with respect to the pump pulse leaves the molecules, which are excited by the pump pulse, time to convert a part of the photoenergy from the pump photon into vibrational energy and relax e.g. into a lower lying electronic state. Thus, the energy gap between the excited and the ion state is enlarged and a higher amount of energy is needed to ionize the molecule. Therefore, the excess energy of the photoelectrons is reduced. In Fig. 4.9 (b) also bands of photoelectron energies higher than the [1,2'] ionization cutoff (marked by a red line) are observable. These result from a [1,3'] photoionization process. However, in the spectral region between 0 and 1.25 eV photoelectron energy, the contribution of [1,3'] ionization is negligible, since 3-photon absorption probabilities scale with the third power of the light intensity and 2-photon absorption quadratically.

In Fig. 4.9 (a) two features can be distinguished: At time zero a very short lived and spectrally broad band ranging from 0 to 3.5 eV is observable. A second band, which is spectrally narrower (0 - 1.3 eV) persists throughout the whole investigated time window of 300 ps. Both features are also found in Fig. 4.9 (b). However, the spectral narrowing between short and long lived bands is not as pronounced as in Fig. 4.9 (a) due to a more confined signature of the short lived (0 - 3 eV) and a broader signature of the persisting band (0 - 2.5 eV). Additionally, both photoelectron bands

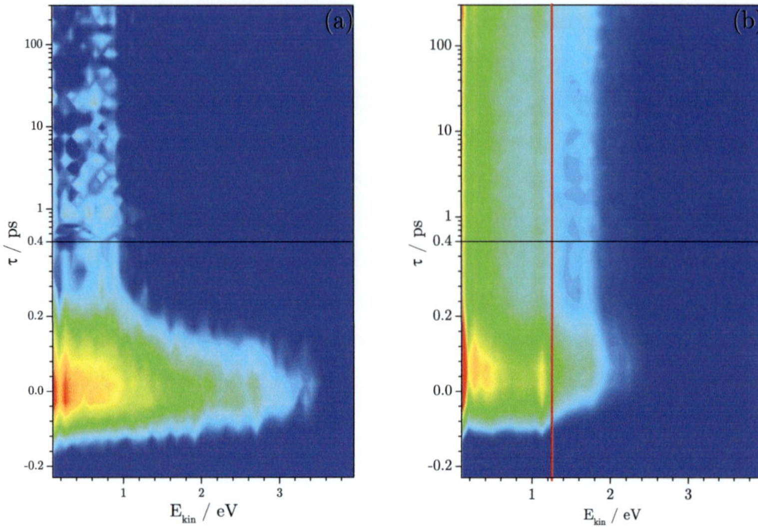

Figure 4.9.: Time-resolved photoelectron spectra of C_5Cl_6 at $\lambda_p = 315$ nm and (a) $\lambda_{pr} = 267$ nm and (b) $\lambda_{pr} = 400$ nm. In the latter spectrum the energy cutoff between [1,2'] and [1,3'] photon ionization at 1.25 eV is marked by a red line.

are more structured than the corresponding ones in Fig. 4.9 (a). The difference can, at least in part, be attributed to the energy cutoff between [1,2'] and [1,3'] photoionization. The intensities of the stable bands suggest that ionization of the underlying species is resonant with both employed λ_{pr}.

The TRPE spectra are cut into slices of $\Delta E = 0.08$ eV and fitted with functions of the type

$$S(E_{kin}, \tau) = g(\tau) \otimes \left[\sum_{i=1}^{2} A_i(E_{kin}) \cdot P_i(\tau) \right], \tag{4.2}$$

in analogy to Eq. 4.1. Here, $P_i(\tau)$ represents the ith exponential depopulation step and $A_i(E_{kin})$ its decay-associated spectrum (DAS). P_1 models the decay of the short lived band and P_2 the evolution of the persisting band.

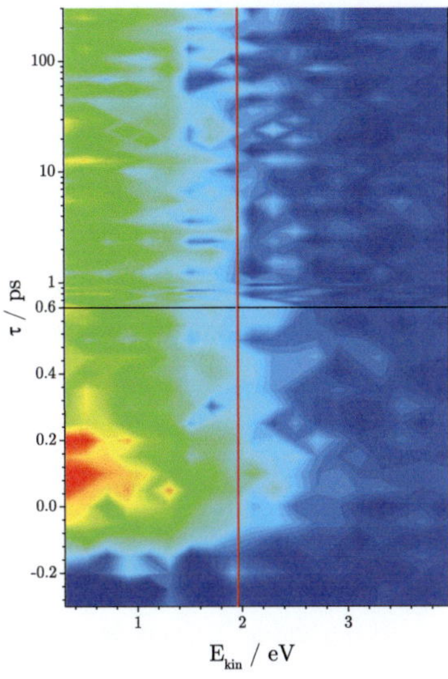

Figure 4.10.: Time-resolved photoelectron spectrum of C_5Br_6 at $\lambda_p = 315$ nm and $\lambda_{pr} = 400$ nm. The energy cutoff between [1,2'] and [1,3'] photon ionization at 1.96 eV is marked by a red line.

The resulting time constants are listed in Tab. 4.4, the DAS are depicted in Fig. A.2 in the appendix A. Errors, upper and lower boundaries of the time constants are given according to the same considerations as in Sect. 4.3. τ_1, which is associated to the short lived band, exhibits a value of < 0.14 ps. Since a decay of the persisting photoelectron band in both TRPE spectra is not observed on the time scale of the experiment, τ_2 is fixed to a very high value, which cannot be defined further.

sample	λ_{pr} / nm	τ_1 / ps	τ_2 / ns
C_5Cl_6	267	< 0.14	> 0.6
C_5Cl_6	400	< 0.14	> 0.6
C_5Br_6	400	0.15 ± 0.14	> 0.6

Table 4.4.: Time constants resulting from global fits of the TRPE spectra in Figs. 4.9 and 4.10 with functions of the type of Eq. 4.2.

Due to a low vapor pressure of C_5Br_6[83] only weak TRPE spectra could be achieved. They are shown in Fig. 4.10. The cutoff between [1,2'] and [1,3'] photon ionization at 1.96 eV (based on an IP of 8.18 eV calculated with DFT) is also marked there. The TRPE spectrum shows a comparable behavior. Like in the case of C_5Cl_6 around time zero a broader, very short lived photoelectron band is observable. An additional, slightly narrower photoelectron band persists throughout the whole investigated time window. Its intensity points to a resonant ionization of the underlying species. The TRPE spectrum is cut into slices of $\Delta E = 0.2$ eV and also fitted with functions of the type of Eq. 4.2. The resulting time constants are listed in Tab. 4.4, the DAS are shown in Fig. A.3 in the appendix A. In agreement with the findings for C_5Cl_6 a very short time constant $\tau_1 = 0.15 \pm 0.14$ ps and a time constant τ_2 with a value beyond the experimentally accessible time scale are found.

4.5. Excited states dynamics on a femtosecond time scale

Obviously, the gas phase dynamics of both C_5Cl_6 and C_5Br_6 show only one single concerted reaction step associated with τ_1. It leads to a species, which is stable on the investigated time scale and seems to be resonant with both λ_{pr} due to the observed band intensities. From the knowledge of the photoinduced reactions in solution[134, 135] one is in favor of assigning the step to homolytic fission of a carbon – halogen bond at the sp^3 position of the five-membered ring. The assignment is strongly supported by the absorption spectra of C_5Cl_5 and C_5Br_5 in solution, which exhibit – besides the temperature-dependent CT absorption band in the visible – additional absorption bands around the employed λ_{pr} of 400 nm.[134, 135] Furthermore, the spectrum of C_5Cl_5 (Fig. 4.11, geometry data in Tab. C.3 in appendix C) calculated by TDDFT shows additional transitions with high oscillator strengths around 267 nm, the other employed λ_{pr}.

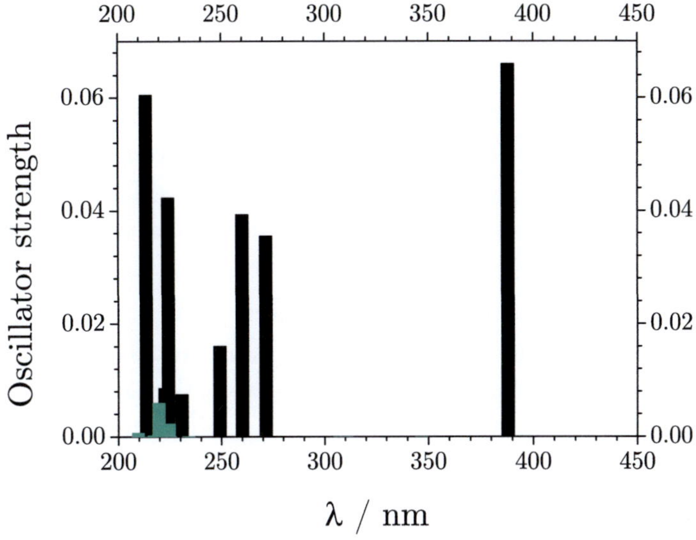

Figure 4.11.: Absorption spectrum of C_5Cl_5 calculated with TDDFT including transitions to states with A' symmetry (black) and A'' symmetry (green). An additional excitation with zero oscillator strength is calculated to appear at $\lambda = 1614$ nm.

C_5Cl_5 is a Jahn-Teller system, which exhibits a degenerate ground state at D_{5h} symmetry and is therefore expected to show symmetry lowering to C_{2v} in the ground state resulting in a so-called umbrella potential for the two symmetry lowering normal modes.[189–192] Nonetheless, EPR spectroscopy in solution gave hints for D_{5h} symmetry.[134] The finding can be explained with the umbrella potential being shallow enough that the first vibrational state is higher in energy than the conical intersection at D_{5h} symmetry. In general, any single-reference method is questionable in the direct vicinity of a point of degeneracy. However, the nearly degenerate electronic states are also found in the TDDFT calculation as a transition with zero oscillator strength at $\lambda = 1614$ nm. Additionally, the first intense transition is found at 400 nm as expected from experimental findings.[134, 135]

Further support for the assignment of the observed photoproducts to C_5X_5 comes from the rather narrow shape of the persisting photoelectron band. The underlying species seems not to be highly vibrationally excited, as otherwise a by far more broadened band would be observed. The fact can be explained, on the one hand, by consumption of a substantial amount of the photoenergy by a photoreaction like the proposed bond breakage and, on the other hand, by the leaving halogen taking along an additional amount of energy as kinetic energy.

Alternatively, a decomposition of the carbon ring is in general possible. For such reactions large amplitude motions of the ring have been observed, which lead to spectral shifts of photoelectron bands.[180] Since such shifts cannot be observed in the present case, a decomposition of the ring is improbable.

The assignment to bond fission is additionally supported by the observed early dynamics of C_5X_6 in solution. The initial decay of TA after time zero, which is connected to τ_1, takes place on the same time scale (see Tabs. 4.2, 4.3 and 4.4). Thus, the time constants τ_1 observed in the gas phase and solution are likely to refer to the same process. The dynamics following the initial decay, therefore, can be assigned to processes involving the two radicals, which originate from the bond fission, and the surrounding solvent. These processes are typically not observable in the gas phase, since a solvent cage, which keeps the two radical species together, is missing.

4.6. Dynamics in solution leading to charge-transfer (CT) complexes

The rise of TA at short λ_{pr} in C_5Br_6 is in good agreement with the known spectra of CT absorption bands in a frozen solvent,[135] which keeps the two photogenerated radicals close together in a joint solvent cage. Since the CT absorption bands are so stable that they could be observed by steady-state spectroscopy, geminate recombination of the radicals does not seem to be an important process, as it typically takes place on a ps time scale.[133,193,194] As melting of the solvent leads to disappearance of the absorption bands, the band intensity seems to be crucially sensitive to the separation of the two radicals. CT complexes of bromine radicals with different solvents are known to exhibit absorption bands in the UV.[130] Therefore, the absorption bands observed here cannot originate from interaction between the bromine radical and the solvent and thus have to be assigned to a geminate CT complex of the two radicals.

In very recent studies the dynamics of benzhydryl cations in solution were investigated.[14,195,196] The cations are, in this case, produced via a heterolytic as well as an homolytic bond dissociation channel, whereas in the latter case the dissociation was followed by a charge transfer on a ps time scale. Why is, in the present case, only a CT absorption observable and not a CT reaction? The fact can be explained by the different electronic structures of, on the one hand, the benzhydryl cation and its precursor, e.g. benzhydryl chloride, and on the other hand, the C_5Br_5 radical and C_5Br_6. Homolytic bond dissociation of benzhydryl chloride leads to a chlorine radical with a high electron affinity and a benzhydryl radical, which can achieve an aromatic configuration by releasing an electron. Therefore, a CT reaction is highly favored. In contrast, by homolytic bond dissociation of C_5Br_6, a bromine radical is generated with high electron affinity and a C_5Br_5 radical, which can become aromatic by receiving an electron and would become antiaromatic by releasing an electron. Thus, both radicals

(a) (b)

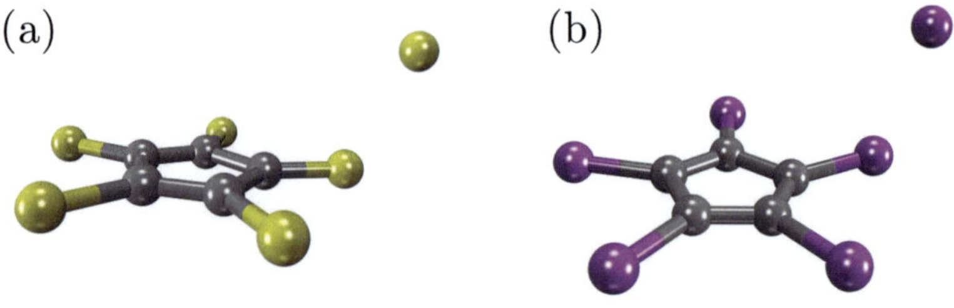

Figure 4.12.: Stuctures of the optimized CT complexes originating from (a) C_5Cl_6 and (b) C_5Br_6.

are only stabilized by reduction and both possible CT reactions are unfavorable.

As opposed to C_5Br_6, CT absorption bands from photolyzed C_5Cl_6 were not observed earlier.[134, 135] The solvent, which was employed at that time, was an apolar hydrocarbon like the solvent cyclohexane. Therefore, the earlier findings are in agreement with the present observations, since in cyclohexane C_5Cl_6 does not exhibit a transient absorption band in the visible. Furthermore, CT complexes of Cl radicals with the solvent exhibit absorptions in the UV.[127, 131] Thus the TA rise on a ps time scale in the other solvents can also be assigned to absorption of a geminate CT complex.

Possible structures for the photolyzed C_5X_6 were investigated by DFT. Only one minimum exhibiting C_s symmetry could be found, where the leaving halogen is loosely bound to another halogen ring substituent. The Cartesian coordinates of the structures are found in Tabs. C.2 and C.5 in appendix C. The halogen – halogen bond is thereby directed perpendicularly to the planar C_5X_5 ring system (see Fig. 4.12). The distance between the chlorine (bromine) atoms is 250 pm (270 pm), which is considerably larger than the bond distance of Cl_2 (Br_2) (199 pm (229 pm)).[6] Moreover, TDDFT calculations reveal that these species exhibit very strong transitions in the spectral region, where the CT absorption band was observed (see Fig. 4.17). It is known that TDDFT tends to poorly describe excitation energies to

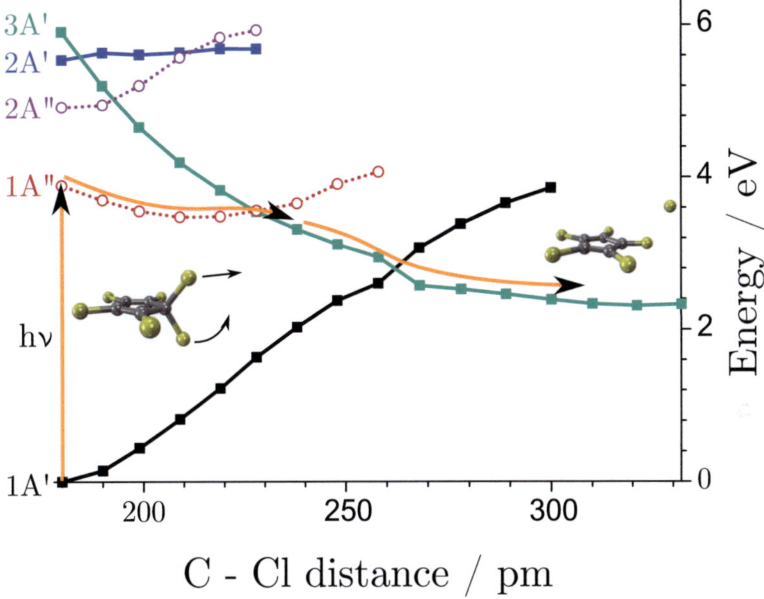

C - Cl distance / pm

Figure 4.13.: Interpolated path between the minimum of C_5Cl_6 and the optimized CT complex minimum calculated with CC2. The states with A' symmetry are labeled by solid lines, the states with A" by dotted lines. The proposed relaxation pattern is marked by arrows. It includes the 1A' ground state (black), the 1A" state (red), which the molecule is excited into, and the 3A' state (green). Intervening states 2A" (violet) and 2A' (blue) are shown not to play a significant role.

CT states[197–199] due to self-interaction.[200] However, the problem can be partially compensated by the use of hybrid functionals like B3LYP as employed in the present case.[201]

An interpolated path between the C_5Cl_6 minimum, and the minimum of the optimized species calculated with CC2 (see Fig. 4.13), reveals a possible relaxation path in the direction of the CT species ($C_5Cl_5\cdots Cl$). In Fig. 4.13 the states with A' symmetry are labeled by solid lines, the states with A" symmetry by dotted lines. The discussion of the interpolated path can be reduced to the ground state (1A') of C_5Cl_6 and the excited

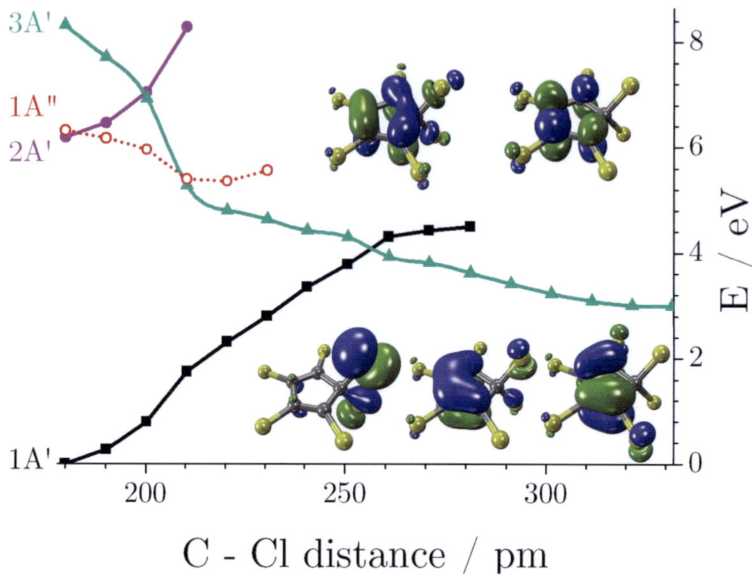

Figure 4.14.: Interpolated path between the minimum of C_5Cl_6 and the optimized CT complex minimum calculated at the SA-4-CASSCF(6,5) level of theory. The states with A' symmetry are labeled by solid lines, the states with A" by dotted lines. The MOs, which compose the active space are visualized. Beginning from the left in the top row they are the LUMO, LUMO+1, HOMO-2, HOMO-1 and the HOMO.

states, which are marked in Fig. 4.3 (a): Photoexcitation from 1A' (black) promotes the molecule to the first excited state with A" symmetry (red). Planarization of the C_5Cl_5 ring and detachment of the Cl atom leads to a moderate lowering of the 1A" state and a substantial lowering of the 3A' state resulting in a crossing. Thus, the population is transferred to the 3A' state, which seems to be at the same time the ground state of the CT complex. The role of the 1A' state, which crosses the 3A' state during the relaxation, cannot be determined by this method due to its single-referential nature.

As discussed in Chap. 2.2, CC2 is known to poorly treat excited states with doubly excited character. Since C_5H_6 exhibits a low lying multiconfigurational state with considerable doubly excited character,[166, 171] it has to be checked if this state is intervening in the picture of the proposed relaxation path. The CC2 results are therefore evaluated by recalculating the same interpolated path for C_5Cl_6 employing SA-4-CASSCF(6,5) (see Fig. 4.14). The active space thereby consists of the 4 π-MOs, two of them occupied, two unoccupied. Additionally, the occupied HOMO-2, which is an anti-bonding linear combination of p-AOs from the chlorine atoms at the sp^3 carbon, is included. These MOs resemble the ones depicted in Fig. 4.3 (a) beside the second unoccupied π-MO, which is also included here. It is averaged over three states with A' symmetry including the ground state and one state with A" symmetry. The 2A' state has a multiconfigurational character with contribution of a doubly excited configuration like the $2A_1$ state of C_5H_6.[166] The 3A' and the 1A" states exhibit the same character as in the CC2 calculation. As a multi-reference method, SA-CASSCF can treat the electronic states qualitatively correctly even at points of degeneracy. However, since it cannot account for dynamic electron correlation (see Chap. 2.2), the relative energies of the electronic states are not exact. Nevertheless, the relaxation path observed by CC2 is qualitatively approved.

Furthermore, it gives an explanation for the substantial lowering of the 3A' state. Upon planarization of the C_5Cl_5 ring one of the p-AOs of the HOMO-2 can interact with the LUMO, which leads to stabilization of the LUMO and destabilization of the HOMO-2 and therefore to an energy lowering of the state. Additionally, according to the CASSCF calculation, the 3A' state is the ground state at the CT complex minimum. Therefore, the 1A' state does not seem to play a role in the relaxation process.

An analogue relaxation path was also calculated for C_5Br_6 with CC2 (see Fig. 4.15). The results are qualitatively comparable to C_5Cl_6. During planarization of the C_5Br_5 ring system and simultaneous elongation of the

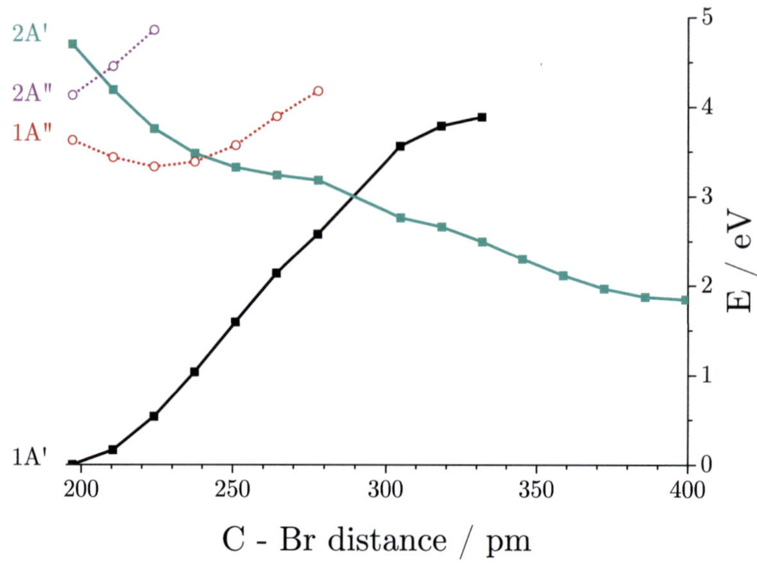

Figure 4.15.: Interpolated path between the minimum of C_5Br_6 and the optimized CT complex minimum calculated with CC2. The states with A' symmetry are labeled by solid lines, the states with A" by dotted lines.

C – Br bond, the 2A' state, which has the same character as the 3A' state in C_5Cl_6, is lowered in energy and crosses with the 1A" state which was excited in the experiments. This seems to be at the same time the ground state of the CT complex.

Despite of the highly simplified picture presented by the interpolated paths, it allows for an explanation of the experimentally observed excited states dynamics, which is summed up by the reaction scheme in Fig. 4.16: Upon relaxation in the 1A" state, photoexcited C_5X_6 reaches a surface crossing between 1A" and 3A' leading to population transfer to 3A', which is associated to the time constant τ_1. In the gas phase (left part of Fig. 4.16) the leaving of the halogen atom is not hindered and the atom takes a major amount of the initial photoenergy along, which is observable by the sharp persisting band in the TRPE spectra. In solution (right part of Fig. 4.16)

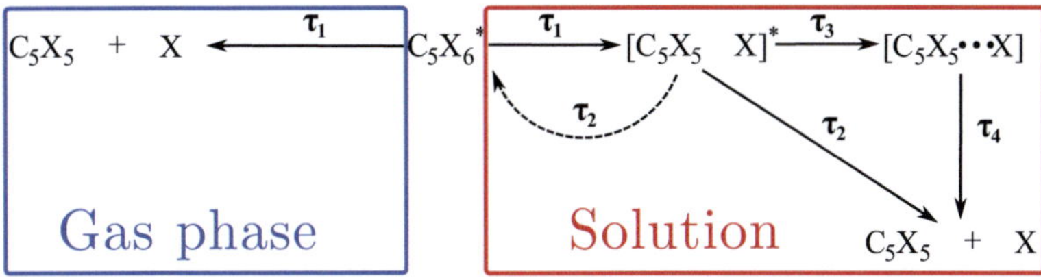

Figure 4.16.: Relaxation scheme for C_5Cl_6 and C_5Br_6 in the gas phase (blue) and solution (red): After excitation C_5X_6 directly dissociates in the gas phase with a time constant τ_1. In solution the same process is also found, but followed by formation of the CT complex associated with the time constant τ_3. Formation of the CT complex competes with direct escape of the halogen radical connected to τ_2. A possible additional contribution to τ_2 is geminate recombination (dotted arrow). Quenching of the CT complex is connected to τ_4.

there is a high probability for the molecules, which constitute the solvent cage, to capture the atom. The system then relaxes on the time scale of τ_3 into the CT minimum ($[C_5X_5\cdots X]$), which was observed by the DFT calculations (see Fig. 4.12). As discussed above, C_5Cl_6 does not show a transient CT absorption band in cyclohexane. Thus, in this case after bond fission, the radicals either immediately recombine (dotted line, τ_2) or the Cl atom escapes the solvent cage (solid line, τ_2). Since the photoproduct in apolar hydrocarbons was identified to be C_5Cl_5 by EPR spectroscopy,[134] there is definitively a nonzero probability for the Cl atom to immediately escape the solvent cage. In general it could be possible that a CT complex is also formed, which absorbs in a totally different spectral region. However, this seems unlikely, as in the other solvents larger effects on the spectrum of the CT band would have to be expected. Thus, in the case of C_5Cl_6 in cyclohexane, τ_2 can essentially be assigned to escape of the Cl radical from the solvent cage. The assignment is further supported by earlier experimental and theoretical studies of radicals and their escape from solvent cages, which takes place on analogue time scales.[133, 193, 194]

In the other samples time constants τ_2 with similar values are found. Therefore, the same assignment is also employed for C_5Br_6 and C_5Cl_6 in the other solvents, whereby recombination/escape are channels competing with formation of the CT complex in these cases.

4.7. Spectral shift and the fate of the charge-transfer (CT) complexes

The CT absorption band can be expected to be sensitive even to little changes in the distance between the radicals. The assignment of the spectral shift of the CT band, which is observed in some solvents, to a change in distance is therefore an obvious interpretation. To support this interpretation, the dissociation coordinate of the $C_5Cl_5\cdots Cl$ complex is scanned by TDDFT. In Fig. 4.17 the energies of the ground state (black) and the excited state with the by far highest oscillator strength (green) are depicted depending on the $Cl - Cl$ bond distance. Additionally, the oscillator strengths for the transition at all calculated distances are shown as bars. The plot gives a qualitative explanation for the observed spectral shift of the CT absorption band. Upon enlargement of the $Cl - Cl$ distance, the transition wavelength shifts towards longer wavelengths and simultaneously the band intensity weakens. Thus, on the time scale of the spectral shift a motion of the halogen radical away from the C_5X_5 radical is observable.

Does this finding indicate a quenching of the CT complex by diffusion? The time scale of the spectral shift can be quantified by the range of τ_3 values in Tab. 4.2 and 4.3. It can be compared to the expectation of simple diffusion. The time dependence of the distance expectation value[202] is

$$\langle d^2 \rangle = 2Dt. \tag{4.3}$$

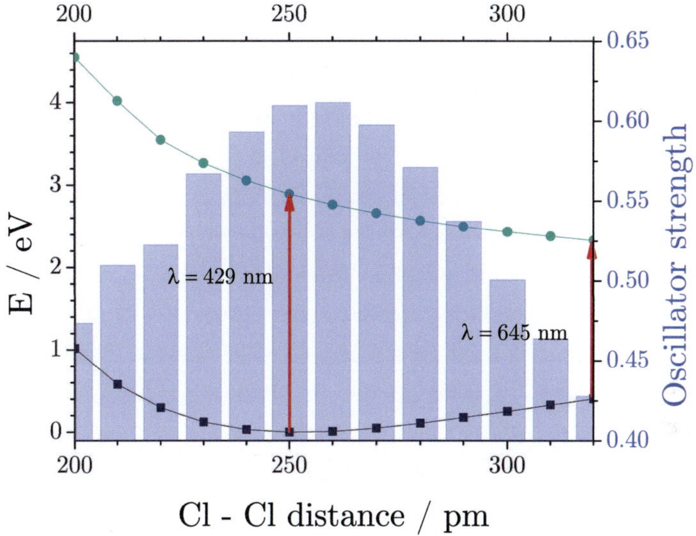

Figure 4.17.: Coordinate scan of the Cl – Cl bond in the $C_5Cl_5\cdots Cl$ CT complex with TDDFT. Connected points: Ground state energies (black) and energies of the excited state with the by far highest oscillator strength in the visible (green). Calculated oscillator strengths are represented by columns.

By employing the Stokes relation for the diffusion coefficient D of a single radical species

$$D = \frac{k_B T}{6\pi\eta r},\qquad(4.4)$$

one can evaluate the time, which is needed by the two radicals to diffuse an average distance $\sqrt{\langle d^2\rangle}$ by

$$t = \frac{\langle d^2\rangle\, 6\pi\eta r_A r_B}{2k_B T\,(r_A + r_B)}.\qquad(4.5)$$

The shape of the radicals is estimated here to be spherical, which is a good approximation for the halogen radical. The C_5X_5 radical has rather an oblate shape, but for an estimation the assumption of a sphere seems to be reasonable. The distance is chosen to be the calculated equilibrium distance in the CT complex, i.e. an estimation is made about the diffusion time for the radicals to cover twice the bond distance (250 pm ($C_5Cl_5\cdots Cl$) vs. 270 pm ($C_5Br_5\cdots Br$)). The radii of the radicals are estimated to be $r_A = 522$ pm (C_5Cl_5) and 560 pm (C_5Br_5). r_B is estimated to be 100 pm (Cl) and 115 pm (Br), which are the covalent radii of the elements.[6] Solvent viscosities are listed in Tab. 4.1. Chloroform exhibits the lowest viscosity ($\eta = 0.56$ mPa s) and trichloroethanol the highest ($\eta = 21$ mPa s).

The values calculated for $C_5Cl_5\cdots Cl$ range between 7 and 260 ps and between 11 and 410 ps for $C_5Br_5\cdots Br$, dependent of the solvent viscosity. The agreement with the range of τ_3 values, which are found in connection to the spectral shift (see Tabs. 4.2 and 4.3), is reasonably good. The shift, therefore, could be due to a diffusional process. Comparison with the time scale of the CT band decay (τ_4), however, shows a difference by an order of magnitude in most cases. A direct connection between spectral shift and quenching mechanism can therefore be excluded. It can be speculated that within the minimum of the CT complex diffusion is possible to a small extent. The time scales of the two processes are however not as different in the case of C_5Cl_6 in trichloroethanol. This finding will be discussed below.

The comparison gives an additional hint for the nature of the CT complex. Apparently, its stabilization is not of purely sterical nature due to the solvent cage. As comparable time scales of CT absorption quenching were found for similar systems like photolyzed bromoform,[133] the present case does not seem to be an exceptional one. As opposed to e.g. the investigation of bromoform, it can be shown here that the lifetime of the CT absorption is not due to a halogen radical, which forms transient CT species with solvent molecules during diffusion, but to a considerably stronger interaction between two individual species.

The quenching, therefore, must be either due to a thermal cleavage reaction, or due to direct reaction of the radicals constituting the CT complex with molecules from the solvent cage. It can only be speculated about the reactivity of the C_5X_5 radical. The most probable reaction is reduction to the anion, which is a stable aromatic. Due to a lower steric demand the reactivity of the halogen radicals, however, can be expected to have a higher influence on the quenching of the CT complex. Cl radicals are known to form CT complexes with halogenated solvents exhibiting absorption bands in the UV.[127, 131] As opposed to Br radicals, which are known to be rather unreactive in solution and have lifetimes in the µs range in many solvents,[125, 126, 128–130, 133] they undergo rapid reaction with solvent molecules under H abstraction.[122, 123, 132] These reactions are assumed to be very fast in the case of aliphatic hydrocarbons and alcohols. The time constant for the reaction between Cl and cyclohexane was measured to be 20 ps.[122] Thus, in the case of $C_5Cl_5\cdots Cl$ the CT complex can, in general, be quenched by direct reaction of Cl radicals with molecules from the solvent cage. However, in the case of C_5Cl_6 in cyclohexane the formation of a CT complex is not observed, which prevents the reaction from being investigated at least in the visible spectral range. Hydrogen abstraction from isopropanol was observed to take place with a time constant of 12 ps.[132] Comparison with the findings regarding the TA decay in isopropanol leads to the conclusion that direct reaction with the solvent cage cannot account for the TA quenching. The same also holds for chloroform. Thus, a thermal decomposition of the CT complex has to be assumed. Reactions of the halogen radicals following the decomposition do not seem to be observable in most solvents.

In contrast, trichloroethanol exhibits a TA decay, which is substantially faster than in the other solvents ($\tau_4 = 220 \pm$ ps). The finding is in good agreement with the estimated diffusion time of 260 ps. Furthermore, the spectral shift, which can be connected to a motion of the radicals away from each other, is on a similar time scale. Therefore, a motion of the radicals away from each other, which is only controlled by diffusion, can be assumed.

The question remains, why a diffusional quenching process is only possible in trichloroethanol. It seems rather unlikely that the CT complex is destabilized to such a great amount by bulk solvent properties, since the properties of isopropanol are generally similar (see Tab. 4.1). A possible reason for the behavior is a lowering of the dissociation barrier, which keeps the radicals together, by a direct hydrogen abstraction channel involving a molecule from the solvent cage. H abstraction could be facilitated for trichloroethanol in comparison to isopropanol by the electron withdrawing effect of the chlorine substitution.

4.8. Conclusion and outlook

The two systems C_5Cl_6 and C_5Br_6 perform photoinduced homolytic bond dissociation both in the gas phase and in solution within $\tau_1 < 100$ fs. Following the bond dissociation in solution, geminate formation of a CT complex is observed on the time scale of several ps, competing with escape of the halogen radical from the solvent cage and geminate recombination of the radicals taking place withing few ps. The competition is thereby highly solvent-dependent in the case of C_5Cl_6. The CT complexes are observed to be surprisingly stable in most of the combinations of solvent and CT complex. Thus, for the first time, the fate of individual CT complexes of halogen radicals can be investigated. Therefore, the results shed new light on the nature of these complexes. In the solvent trichloroethanol the $C_5Cl_5\cdots Cl$ CT complex is observed to be less stable by an order of magnitude than in the other solvents, which can be regarded as a hint for a direct H abstraction reaction of the Cl radical with a molecule from the solvent cage.

The excited states dynamics of the CT complex are a potentially interesting topic for further time-resolved investigations. Since in the excited CT state the halogen radical is expected to obtain a positive charge, the resulting species should be highly reactive towards either recombination to C_5X_6

or full dissociation accompanied or followed by reaction of the halogen with the solvent environment. An additional issue for further investigations is the fate of the C_5Cl_5 radical in the solvent cyclohexane. In this context not only its lifetime in solution could be investigated but especially its electronic structure. Since its ground state is expected to be degenerate, it should show anomalous values of initial anisotropy[43, 203–205] and could therefore be an interesting model system.

5. Early steps in polymerization initiation by type I photoinitiators

5.1. Introduction

The work presented in this chapter has been conducted in collaboration with Dominik Voll in the group of Christopher Barner-Kowollik, Institute for Chemical Technology and Polymer Chemistry, Karlsruhe Institute of Technology, and has been published in part.[206]

Triplet photoinitiators have recently gained a lot of research interest[207–209] due to their applicability in lithography,[210,211] photocuring,[212] biomaterials[213,214] and dental restorative materials.[215,216] They can be sorted into two types depending on the nature of their reaction with monomer molecules:[217] Type I photoinitiators decompose from the triplet state into radicals by α-cleavage.[218] Type II photoinitiators produce radicals from the triplet state by energy transfer, electron transfer or hydrogen transfer.[209,219–223] The photoinitiators investigated in the present chapter all belong to type I. They share the basic structure of two six-membered aromatic rings, which are bridged by a substituted C_2 unit (see Fig. 5.1, green frame). Radicals (blue frame) are produced by homolytic photolysis of the carbon bridge. Two of the three photoinitiators, which are comparatively investigated, exhibit very similar structures. Benzoin (2-hydroxy-1,2-diphenylethanone, Bz) and 2,4,6-trimethylbenzoin (2-hydroxy-1-mesityl-2-phenylethanone, TMB) are both aromatic α-hydroxy-ketones, which differ only by the aromatic substituent on the carbonyl group. The third photoinitiator mesitil (1,2-bis(2,4,6-trimethylphenyl)-1,2-ethanedione, Me) is a diketone. The main structural difference to the other two photoinitiators

benzoin   2,4,6-trimethylbenzoin   mesitil

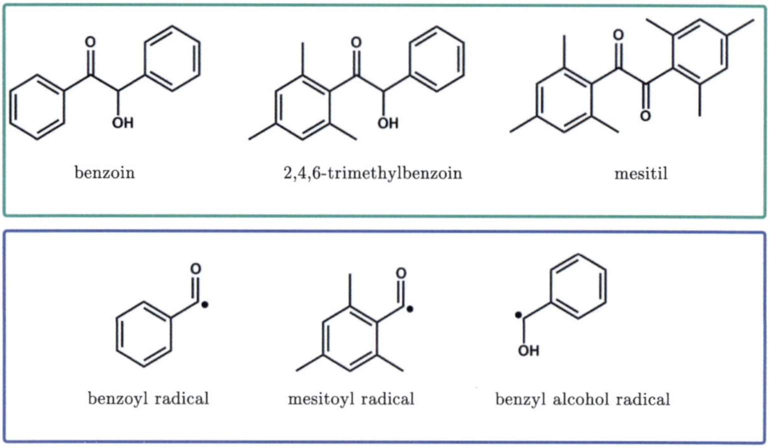

benzoyl radical   mesitoyl radical   benzyl alcohol radical

Figure 5.1.: The investigated photoinitiators (green frame) and all possible radicals (blue frame), which can be produced from the photoinitiators by irradiation.

is the absence of an sp^3-substituted carbon in the bridge between the aromatic rings. However, steric effects induced by the methyl substituents at the aromatic rings are expected to prevent the molecule from being planar, which could have a similar effect on the resonance between the rings as the sp^3-carbon in the other two photoinitiators.

These three photoinitiators were comparably investigated earlier with regard to their relative polymerization initiation efficiency by pulsed laser polymerization (PLP) at $\lambda = 351$ nm with subsequent product analysis by size exclusion chromatography / electrospray ionization mass spectrometry (SEC/ESI-MS).[7] The relative efficiencies were extracted by exploring the incorporation distribution of the three possible starter radicals generated from the photoinitiators (see the blue frame of Fig. 5.1) into polymers. Thereby, all experiments were conducted under the same conditions including photoinitiator concentration, photolysis laser intensity and wavelength. The surprising result of the comparison was that the incorporation rate was not only dependent on the chemical nature of the specific starter radical

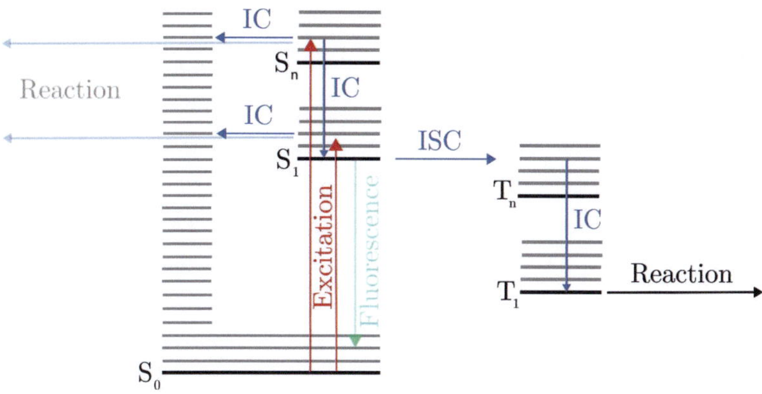

Figure 5.2.: Jabłoński-type diagram of the excited state processes following photoexcitation of a photoinitiator molecule, adapted from Fig. 1.1. The processes, which are relevant in this chapter are highlighted. They are ultrafast radiationless IC between excited states, IC between excited states and S_0, and ISC into the triplet manifold followed by radical formation from T_1.

(blue frame of Fig. 5.1) but also on its origin, i.e. on the triplet photoinitiator (green frame of Fig. 5.1). Benzoyl radicals originating from Bz were observed to exhibit an incorporation probability between 2.4 and 3 times higher than mesitoyl radicals originating from TMB, depending on the employed monomer. In contrast to this finding mesitoyl radicals from Me were 8.6 times less often incorporated than the benzoyl radicals from Bz. The reasons especially for the latter finding seem to lie in the different excited states properties and dynamics of the three photoinitiators.

The possible processes following photoexcitation of a photoinitiator are discussed in the following in more detail with the help of the Jabłoński-type diagram in Fig. 5.2, which is adapted from Fig. 1.1. The processes relevant for the relaxation dynamics, which are described in this chapter, are highlighted. Photoexcitation takes place from S_0 into S_1 or a higher excited singlet state (S_n). Rapid radiationless relaxation by IC into S_0 is not desired in the case of photoinitiators, but cannot be ruled out a priori, especially after excitation into higher excited singlet states. The reason is the growing density of states while approaching the ionization threshold and the thereby growing probability for points of (near) degeneracy between electronic states. Degeneracies, also known as conical intersections, offer highly efficient radiationless channels for depopulation of an electronic state.

Another exit channel from S_1 is ISC, in most cases into a higher triplet state (T_n), followed by IC into T_1. This exit channel is the desired one for triplet photoinitiators, because radical formation takes place from T_1. The efficiency of ISC is governed by two aspects, the character of the involved states and the energy gap between them. The influence of the state characters is well summarized by the El Sayed selection rule, which predicts efficient ISC only between states, where one of them has $n\pi^*$ character and the other $\pi\pi^*$ character.[224] Furthermore, the ISC efficiency is also crucially sensitive to the energy gap between the involved states. In particular, ISC efficiency exponentially declines with a growing energy gap.[225]

A potential further exit channel for S_1 population is fluorescence from the S_1 minimum. This process is also unwanted in the case of photoinitiators. Suitable molecules, therefore, often can be identified by low transition dipole strengths between S_1 and S_0, which also leads to a weak first absorption band in the spectrum[226–228] (see also the spectra in Fig. 5.3).

In summary, an efficient photoinitiator is characterized by an efficient ISC into the triplet manifold and weak further radiative and nonradiative relaxation channels into S_0, to transfer as much population as possible to T_1.

Bz was subject of several theoretical[225] and experimental[218,229,230] studies. It is known to rapidly undergo ISC after photoexcitation. The T_1 state is known to be depopulated on a ps time scale. However, due to poor time resolution the exact mechanism and time scale of ISC in Bz could so far not be investigated. The excited state properties of TMB and Me are hitherto investigated neither experimentally nor theoretically.

Two additional important processes in polymerization initiation are not included in the Jabłoński-type diagram in Fig. 5.2: geminate recombination of the produced radicals as discussed in Chaps. 1 and 4 and reaction of the radicals with the monomer starting the polymerization chain reaction. The first reaction can be neglected in the present case. In contrast to radicals, which are generated from a singlet state like the species in Chap. 4, geminate radicals from a triplet state have to undergo a spin flip before being able to recombine. Since such a spin flip is a forbidden process, it presumably takes place on a µs time scale and is, therefore, not observable by the methods employed here.

For an investigation of the reaction between photogenerated radicals and monomer molecules the photoinitiators would have to be examined in monomer solution by transient absorption spectroscopy. Even in a flow cell system polymerization would be started continuously by the pump pulse. The presence of the generated polymer in the flow cell system would make the realization of the experiment extremely difficult. However, as will be shown later, the influence of the reaction between radicals and monomer molecules on the relative incorporation efficiencies can be estimated to a sufficient accuracy.

Thus, it seems to be appropriate to divide the investigation into four sections: (a) Investigation of the static absorption spectra of the three photoinitiators to get an information on the excitation probability, which will be discussed in Sect. 5.2, (b) time resolved investigation of the excited states dynamics, to study the population evolution from the initially excited state to T_1 and further to radical generation, which will be inspected in Sect. 5.3, (c) discussion of the underlying reasons for both the observed static and dynamic excited states behavior with the help of quantum chemical calculations, which will be pointed out in Sect. 5.4, and (d) comparison of the obtained results with the efficiency relations from the earlier PLP-experiments, to get some information on the processes following radical generation and leading to polymer chain reaction, which cannot be investigated by the present time-resolved method. The latter will be discussed in Sect. 5.5.

5.2. Absorption spectra of the photoinitiators

The wavelength-dependent extinction coefficients of Bz, TMB and Me are depicted in Fig. 5.3. As expected from the similar structure, Bz and TMB show also comparable spectra. The maximum of the first absorption band of Bz (green) at 320 nm exhibits a low extinction coefficient of $3.01 \cdot 10^2$ l/(mol·cm). The second absorption band of Bz (not shown in Fig. 5.3, see Fig. 5.8 (a)) features a maximum at 247 nm with an extinction coefficient of $1.32 \cdot 10^4$ l/(mol·cm) and a shoulder at its red flank. The first absorption band of TMB (blue) is also rather weak and slightly blue-shifted with respect to the corresponding band of Bz. Since the second absorption band of TMB is red-shifted (maximum: 254 nm) with respect to Bz, the first absorption band of TMB appears only as a shoulder of the second band. The additional shoulder observed in the spectrum of Bz is not found. The maximum extinction coefficient of the second absorption band ($3.10 \cdot 10^3$ l/(mol·cm)) is considerably lower than that of Bz. The spectrum of Me (red) is distinctively red-shifted with respect to the other two compounds. It exhibits a very weak absorption band with two maxima ($5.06 \cdot 10^1$ l/(mol·cm) at 464 nm and $5.30 \cdot 10^1$ l/(mol·cm) at 493 nm). The second absorption band centered at 286 nm is highly structured and exhibits a maximum extinction coefficient of $2.44 \cdot 10^3$ l/(mol·cm).

Inspecting the extinction coefficients at the excitation wavelength (351 nm) of the earlier PLP-experiments (denoted by λ_p in Fig. 5.3), one can observe highly different values. Me is excited at the red flank of its second absorption band and, therefore, exhibits a comparably high value of $1.08 \cdot 10^3$ l/(mol·cm). Bz is excited at the red flank of its first absorption maximum. The value of its extinction coefficient is, therefore, considerably lower ($6.4 \cdot 10^1$ l/(mol·cm)). Due to excitation at the red edge of its first absorption band, TMB exhibits the lowest value of $1.2 \cdot 10^1$ l/(mol·cm).

One can now compare the above findings to the observed incorporation efficiency relations of the radicals originating from the three photoinitiators.

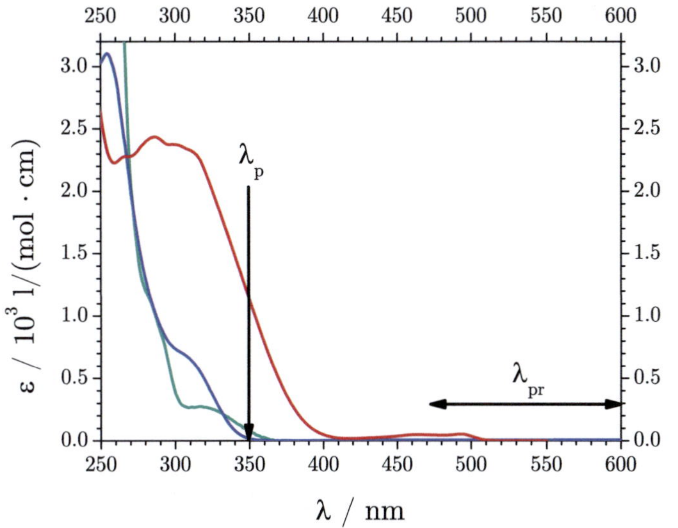

Figure 5.3.: Absorption spectra of Bz (green), TMB (blue) and Me (red). Additionally, λ_p and the λ_{pr} range are marked.

As mentioned above, the relations were measured under identical experimental conditions including concentration, laser intensity, and wavelength. However, as can be estimated by the huge differences in extinction coefficients, the number of photoinitiator molecules, which were actually excited, was not the same. Obviously, the amount of excited Me is by far the highest and the amount of excited TMB the lowest. In a second step the earlier observed efficiency relations can be normalized to the relative amounts of excited molecules. The new relations reflect the relative probabilities of single excited Bz, TMB or Me molecules to produce radicals, which are incorporated into polymer chains. They differ highly from the original relations: The relation between Bz and TMB is nearly inverted with an initiation probability of TMB which is by factors of 1.8 to 2.2 higher than the probability of Bz. The relation between Bz and Me is enlarged by approximately a factor of 17 resulting in an initiation probability of Bz which is higher by a factor of 144 in comparison to Me. Since Bz and TMB exhibit a comparable structure and absorption spectrum, the observed incorporation relation could be attributed to steric effects in the polymerization initiation reaction of the benzoyl and mesitoyl radicals. However, the completely different numbers in the relation between Bz and Me lays additional stress on the earlier assumption that the reasons for the observed efficiency pattern have to be found in the excited states properties of the three molecules.

5.3. Time-resolved investigation of the excited states dynamics

Transient absorption (TA) traces of Bz, TMB, and Me in methanol solution at $\lambda_p = 351$ nm and $\lambda_{pr} = 470$, 500, and 600 nm are shown in Figs. 5.4 and 5.5 together with a global fit analysis employing the same type of functions as in Chap. 4.3 (Eq. 4.1). The motives for the choice of exponential decay functions are also the same. In the present case, a sum of three exponential decay functions $P_i(\tau)$ was sufficient to model the experimental data. A sum of 3 exponential functions yielded a nontrivial improvement on the χ^2 values with respect to 2, further addition of exponential functions did not. Furthermore, a comparison of the TA traces of all three molecules at $\lambda_{pr} = 500$ nm is shown in Fig. 5.6. The range of λ_{pr} is chosen to be outside of spectral regions with high ground state absorption of the photoinitiators. Me exhibits a weak absorption band in the region of $\lambda_{pr} = 470$ nm. However, the extinction coefficients are lower than the extinction coefficient at λ_p by at least a factor of 22. Thus, a possible contribution of ground state bleaching to the TA intensity can be neglected. At $\lambda_{pr} = 600$ nm no TA signal could be found in the case of Me.

The time constants and relative amplitudes resulting from the fits are listed in Tabs. 5.1 and 5.2. Like in Chap. 4 the errors for amplitude and

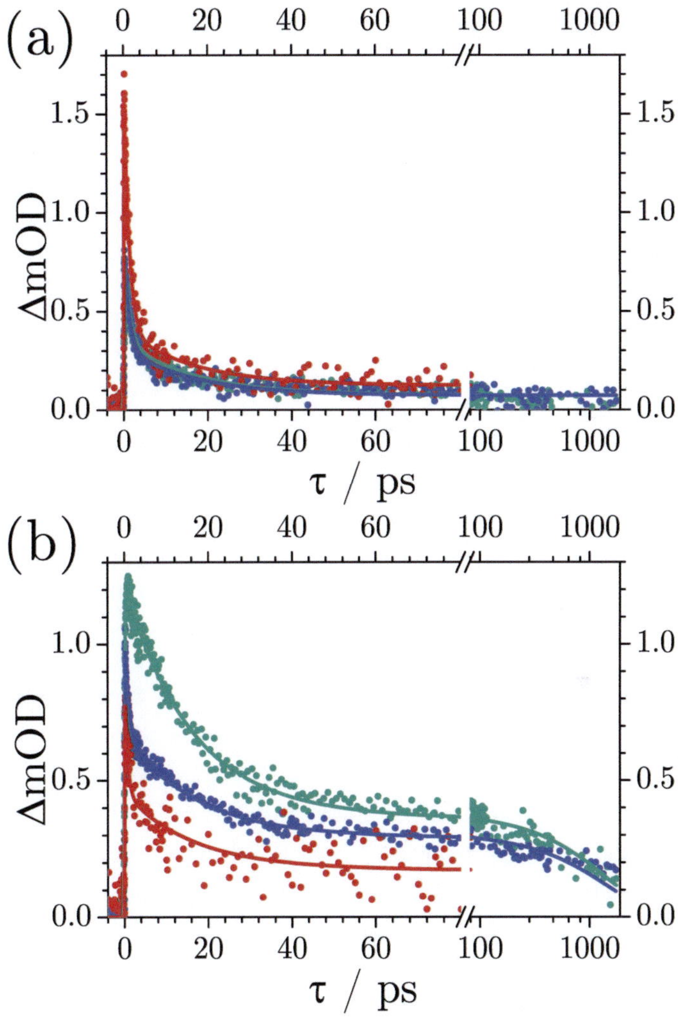

Figure 5.4.: TA traces (dots) of (a) Bz and (b) TMB in methanol at $\lambda_p = 351$ nm and $\lambda_{pr} = 470$ nm (green), 500 nm (blue) and 600 nm (red) together with global fit analysis (lines).

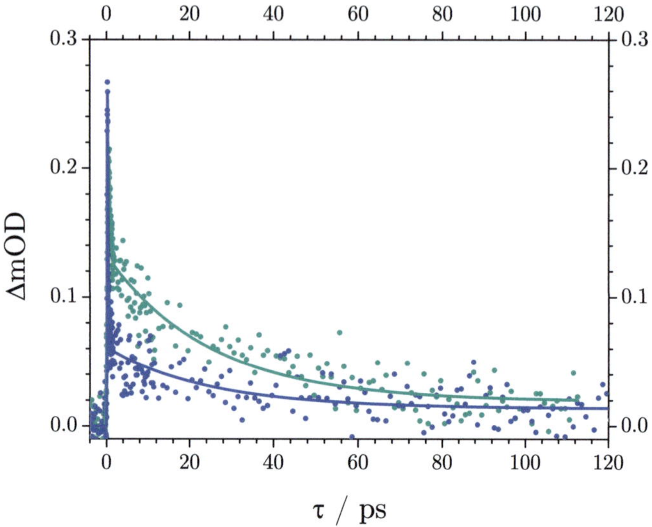

Figure 5.5.: TA traces (dots) of Me in methanol at $\lambda_p = 351$ nm and $\lambda_{pr} = 470$ nm (green) and 500 nm (blue) together with global fit analysis (lines). No TA was found at $\lambda_{pr} = 600$ nm.

time constant values are extracted from the fitting routine. For time constant values outside the experimental time window lower boundaries are given.

Two features are found in all presented TA traces (see also Fig. 5.6): The TA intensity decays in a biexponential manner on a ps time scale to a distinct value. The remaining TA decays on a ns time scale, which is beyond the experimentally accessible time window of 1.6 ns. In the case of Me the decay on a ns time scale was not detectable throughout the whole

Photoinitiator	τ_1 / ps	τ_2 / ps	τ_3 / ns
Bz	1.2 ± 0.2	14.3 ± 0.9	> 3.2
TMB	0.7 ± 0.2	15.9 ± 0.4	> 3.2
Me	0.2 ± 0.2	25 ± 2	> 3.2

Table 5.1.: Time constants resulting from global fit analysis of the TA data.

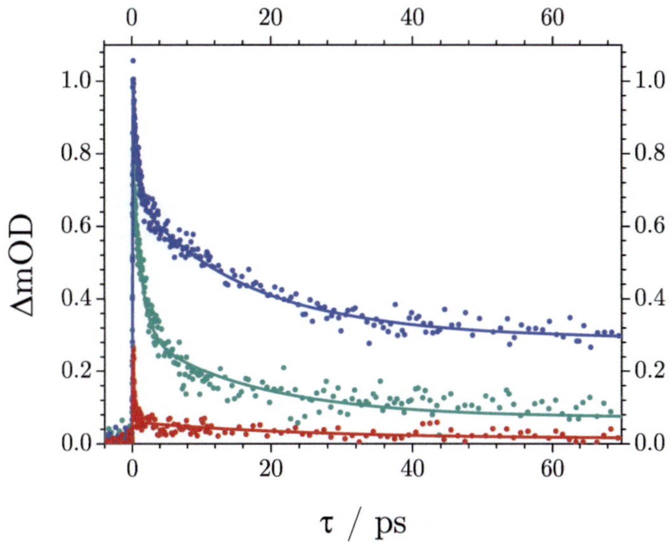

Figure 5.6.: Comparison of the TA traces (dots) together with global fit analysis (lines) of Bz (green) TMB (blue) and Me (red) in methanol at λ_p = 351 nm and λ_{pr} = 500 nm.

time window due to a very low intensity. The absolute TA intensity values of Bz and TMB around time zero are similar, especially in the case of λ_{pr} = 500 nm. The corresponding intensities of Me are lower by approximately a factor of 4. Bz and TMB exhibit very similar time constants for the biexponential decay on a ps time scale (τ_1 = 1.2 $\pm$ 0.2 ps and τ_2 = 14.3 $\pm$ 0.9 ps for Bz vs. τ_1 = 0.7 $\pm$ 0.2 ps and τ_2 = 15.9 $\pm$ 0.4 ps for TMB), whereas the values found for Me are considerably different (τ_1 = 0.2 $\pm$ 0.2 ps and τ_2 = 25 $\pm$ 2 ps). The value of τ_1 thereby approaches the time resolution of the experiment. Thus, the underlying process is most likely faster than the value suggests. τ_2 is by a factor of approximately 1.7 higher than in Bz and TMB. The values of the respective third time constants resulting from the fits are beyond the investigated time window and therefore set to a high value, which cannot be defined further.

The long-lived TA intensity is the highest in the case of TMB and the lowest in the case of Me. A large dependence of TA can only be found for TMB. For comparison, the relative amplitudes from the fits are collected in Tab. 5.2. The relative errors of the amplitudes are calculated from the standard deviations of the values yielded by the fitting routine. For Bz the first component at λ_{pr} = 500 nm has the highest contribution to the TA (see also Fig. 5.6). The second and third components have smaller contributions by factors of close to 2 and 10, respectively. In TMB all three components nearly equally contribute to the TA. In Me the first component has the by far highest contribution. The second component is lower by a factor of 7 and the third by a factor of 30.

Earlier studies of Bz found values for the triplet lifetime of 20 ps and 830 ps,[218,229] where only the first value was obtained from time-resolved measurements and additionally was equal to the time resolution of the experiment.[229] A time constant for the ISC was only estimated. In similar systems time constants for the T_1 depopulation by α-cleavage were also observed to exhibit values in the order of ps.[231–233] Hence, the assignment of τ_2 to depopulation of T_1 by α-cleavage seems to be justified. Accordingly, the TA intensity, which decays on a ns time scale, has to be assigned to transient absorption of the radicals produced by α-cleavage. Due to the

λ_{pr} / nm	Photoinitiator	$A_{1,rel}$	$A_{2,rel}$	$A_{3,rel}$
	Bz	0.39 ± 0.06	0.52 ± 0.03	0.08 ± 0.01
470	TMB	0.08 ± 0.05	0.66 ± 0.01	0.27 ± 0.01
	Me	0.66 ± 0.07	0.29 ± 0.01	0.05 ± 0.01
	Bz	0.60 ± 0.03	0.32 ± 0.02	0.07 ± 0.01
500	TMB	0.33 ± 0.03	0.40 ± 0.01	0.28 ± 0.01
	Me	0.85 ± 0.04	0.12 ± 0.01	0.03 ± 0.01
600	Bz	0.76 ± 0.01	0.17 ± 0.01	0.07 ± 0.01
	TMB	0.50 ± 0.05	0.30 ± 0.01	0.20 ± 0.01

Table 5.2.: Relative amplitudes of the TA of Bz, TMB and Me derived from global fit analysis.

absence of monomer molecules, the radicals are presumably quenched by geminate recombination on a time scale of several μs. As already discussed, the time scale of geminate recombination seems to be too long if compared to analogue reactions discussed in Chap. 4. However, as opposed to geminate radicals generated from a singlet state, radicals from a triplet state have to perform a spin flip to be able to recombine. Side processes like reaction with the solvent methanol seem not to play an important role, since a high photostability of the samples was observed (see Chap. 3.5.2).

The time constant τ_1 is most likely connected to processes, which lead to depopulation of the initially excited singlet state. The value of τ_1 suggests that ISC is not the only process contributing to depopulation of the excited state. Typical ISC time constants are in the range between several ps and μs.[234] Although even ISC time constants below 100 fs have been reported for aromatic systems, these findings could be connected to special anomalies in the electronic structure evoked by an NO_2 group,[235–237] which is not present in the molecules investigated here. However, ISC time constants of 2-3 ps have already been observed for similar systems undergoing α-cleavage.[238] The time constant observed here therefore results most likely from a convolution of two depopulation processes: ISC, and a radiationless relaxation channel, according to $\tau_1 = \frac{\tau_{IC}\tau_{ISC}}{\tau_{IC}+\tau_{ISC}}$. It, therefore, only sets a lower limit to the real ISC time constant.

The high similarity between the TA traces of Bz and TMB suggests an analogous interpretation of the data. Thus, τ_1, τ_2 and τ_3 are assigned to depopulation of the initially excited singlet state, T_1 depopulation, and radical lifetimes, respectively.

In the case of Me the relaxation process associated with τ_1 is considerably faster than in the other two molecules. Applying the same considerations of the two relaxation processes IC and ISC competing for the population of the initially excited state, in the case of Me the dominant process is definitely IC. Thus, ISC is only a minor exit channel from S_1. The time constants τ_2 and τ_3 can also be assigned to depopulation of T_1 and the lifetime of the radicals. The observed intensities agree well with the assignment. The fact that the initial TA intensity is lower by a factor of 4 than in the other molecules can be rationalized by a convolution of the decay with the instrument response function around time zero. The weakness of the TA values observed at longer delay times fits to the assumption that ISC is a minor channel, i.e. the by far major part of population is directly transferred back to S_0 and, therefore, does not contribute to TA anymore.

5.4. Density functional theory (DFT) studies

What are the reasons of the observed different behavior of Bz and TMB vs. Me? For an answer to that question the three photoinitiators are investigated by DFT (for details see Chap. 3.6). The observed minimum energy structures are depicted in Fig. 5.7. Two ground state minima are found for Bz. Their energies are collected in Tab. 5.3, their coordinates in Tabs. C.6 and C.7 in appendix C. The minimum labeled as Geometry 1 exhibits an intramolecular hydrogen bond between the OH group and the neighboring keto group. The minimum labeled as Geometry 2 mainly differs from Geometry 1 by a rotation around the $C-C$ bond, which connects the OH and the keto group. Formation of an intramolecular hydrogen bond

Molecule	Geometry	E [hartree]	E_{corr} [hartree]	ΔE $\left[\frac{kJ}{mol}\right]$
Bz	1	-690.8328	-690.6104	21.8
	2	-690.8230	-690.6024	
TMB	1	-808.6958	-808.3939	24.7
	2	-808.6864	-808.3845	
Me	1	-925.3540	-924.9923	4.7
	2	-925.3513	-924.9905	

Table 5.3.: Energies (E) and zero point corrected energies (E_{corr}) of the optimized geometries of Bz, TMB and Me calculated with DFT. Additionally, the energy differences between the minima (ΔE) are listed.

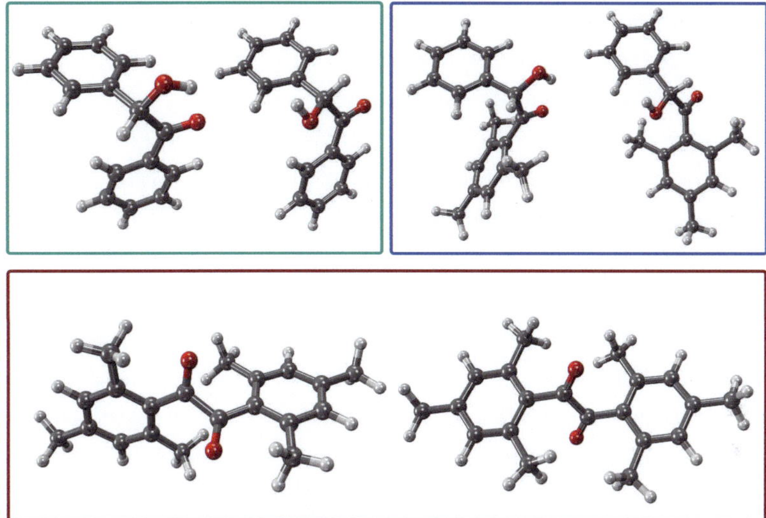

Figure 5.7.: Optimized geometries of Bz (green frame), TMB (blue frame) and Me (red frame). Both Bz and TMB feature ground state minima with an intramolecular hydrogen bond (Geometry 1, left) and without a hydrogen bond (Geometry 2, right). Me exhibits minima, where the planes of the two mesityl rings are oriented perpendicular (Geometry 1, left) and parallel (Geometry 2, right) to each other.

is thus not possible in this geometry. The energies of the two minima differ by 21.8 kJ/mol, which represents typical hydrogen bond energies between methanol molecules.[239] In methanol solution Geometry 2 is energetically stabilized, since the missing intramolecular hydrogen bond can be compensated by intermolecular hydrogen bonds with solvent molecules. Hence, the geometries are presumably closer in energy in methanol solution as the calculations suggest. Furthermore, there is clear experimental evidence, that Geometry 2 is the most stable and active species for radical generation in methanol solution.[229] Hence, Geometry 2 is regarded to be the relevant geometry for further consideration.

TMB shows similar ground state minimum geometries as Bz (for coordinates see Tabs. C.8 and C.9). Geometry 1 also exhibits an intramolecular hydrogen bond, whereas this bond is absent in Geometry 2. Additionally, the energy difference between the geometries is comparable (24.7 kJ/mol). Based on the same considerations as for Bz, Geometry 2 is also regarded as the relevant ground state minimum.

For Me two ground state minima can also be found (for coordinates see Tabs. C.10 and C.11). At Geometry 1 the planes of the two mesityl rings are oriented perpendicularly to each other resulting in a C_2 symmetry, at Geometry 2 the planes are parallel. As the molecule is not able to form intramolecular hydrogen bonds, the minimum lower in energy, Geometry 1, is assigned to be the relevant geometry for the following considerations.

At the relevant geometries of the three photoinitiators excitation energies and oscillator strengths are calculated with TDDFT (for details see Chap. 3.6). Additionally, triplet excitations are calculated. The results are shown in Figs. 5.8 and 5.9 and additionally listed in Tab. 5.4. The agreement of calculated transitions with the experimentally obtained spectra is reasonably good, if one takes into account that no solvent effect was included. Only the calculated transition energy to S_2 in TMB (see Fig. 5.8 (b)) seems to be clearly underestimated, which probably is an effect of the neglect of solvation. Since the effect by methanol is distinct due to hydrogen bonding, inclusion of the solvent in the calculations is desirable. However, less computationally demanding continuum models like the conductor-like screening model COSMO[240] do not describe the solvent as individual molecules, but as a dielectric continuum. Therefore, effects like hydrogen bonding are not included. The computational demand for calculation of a system consisting of the photoinitiator and a solvent cage of methanol molecules to account for hydrogen bonding is by far too high.

As the spectrum in Fig. 5.8 (a) shows, excitation of Bz at $\lambda_p = 351$ nm exclusively leads to population of the first excited singlet state at the Franck-Condon geometry. It has the character of a LUMO$\leftarrow$HOMO excitation.

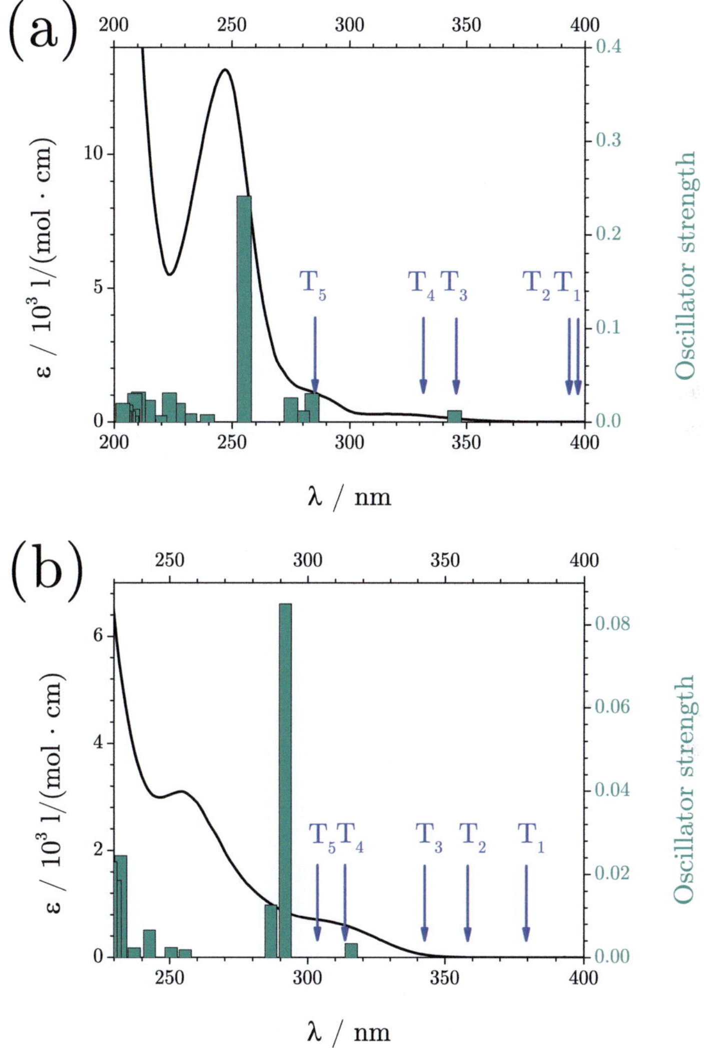

Figure 5.8.: Comparison of the experimental absorption spectra of (a) Bz and (b) TMB with transitions to singlet (bars) and triplet states (marked by arrows) calculated with TDDFT.

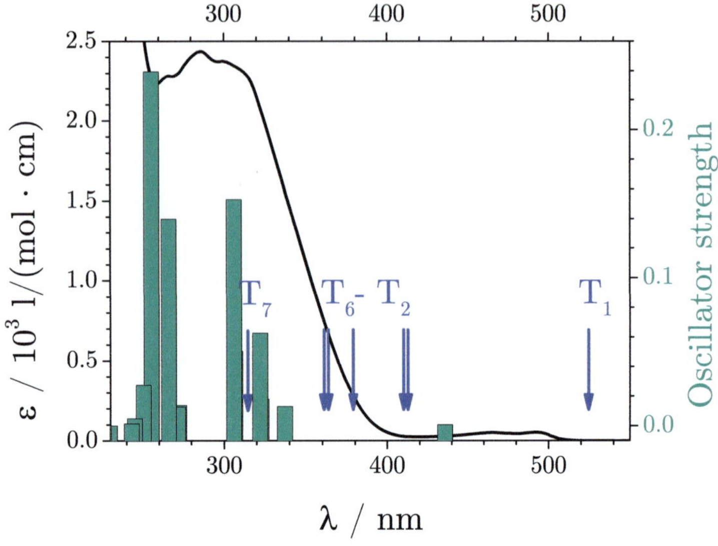

Figure 5.9.: Comparison of the experimental absorption spectrum of Me with transitions to singlet (bars) and triplet states (marked by arrows) calculated with TDDFT.

The involved MOs are visualized in the green frame of Fig. 5.10. As can be seen from the electron density distribution, both MOs exhibit the character of a π-orbital at one of the aromatic rings and of a lone pair with suitable orientation at the carbonyl oxygen. Promotion of an electron from the HOMO to the LUMO of Bz therefore leads to an electron transfer from one aromatic π-system over the isolating sp^3 carbon to the other aromatic π-system. Moreover, one can attribute the low oscillator strength, observed both in the calculated and the experimentally obtained spectra, to the low spacial overlap between the involved MOs.

In the case of TMB excitation at $\lambda_p = 351$ nm also can be exclusively assigned to population of the first excited singlet state (see Fig. 5.8 (b)), whereat – like in the experimental spectrum – the excitation is carried out considerably red-shifted with respect to the transition. The first excited

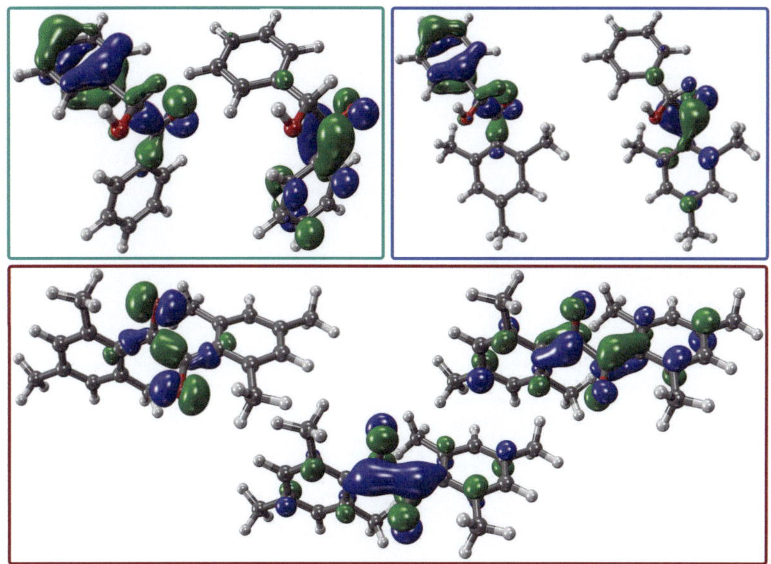

Figure 5.10.: Molecular orbitals involved in the transitions to the respective first electronically excited states of Bz, TMB and Me, calculated with TDDFT: HOMO (left) and LUMO (right) of Bz (green frame), HOMO-2 (left) and LUMO (right) of TMB (blue frame) and HOMO (left), LUMO (middle) and LUMO+1 (right) of Me (red frame).

singlet state has mainly the character of a LUMO←HOMO-2 excitation. However, the character of the involved MOs is the same as in Bz (see the blue frame in Fig. 5.10). Therefore, the same considerations as for Bz also apply to TMB.

The first singlet transition of Me is also governed by a LUMO←HOMO excitation. The involved MOs are depicted in the red frame of Fig. 5.10 (left and middle). It can be seen that despite the earlier considerations about the highly twisted relative orientation of the two aromatic rings, the MOs show considerable resonance interaction between them over the $C-C$ bridge. Thus, the steric effect of the methyl substituents at the aromatic rings cannot substitute the isolating effect of an sp^3 hybridized carbon

Molecule	Singlet state	E [eV]	Triplet state	E [eV]
	S_1	3.59	T_1	3.13
	S_2	4.37	T_2	3.17
Bz	S_3	4.43	T_3	3.58
	S_4	4.51	T_4	3.74
	S_5	4.86	T_5	4.36
	S_1	3.93	T_1	3.27
	S_2	4.25	T_2	3.46
TMB	S_3	4.33	T_3	3.62
	S_4	4.85	T_4	3.96
	S_5	4.95	T_5	4.09
	S_1	2.85	T_1	2.36
	S_2	3.68	T_2	3.01
	S_3	3.84	T_3	3.02
Me	S_4	3.85	T_4	3.28
	S_5	4.04	T_5	3.41
	S_6	4.05	T_6	3.43
	S_7	4.55	T_7	3.98

Table 5.4.: Excitation energies to the lowest singlet and triplet states of Bz, TMB and Me calculated by TDDFT.

atom within the bridge. Hence, the conjugated π-system is considerably larger than in the other two molecules, which is probably the main reason for the red-shift of the whole spectrum with respect to Bz and TMB (see Fig. 5.3). In contrast to Bz and TMB, the HOMO features the character of oxygen lone-pairs, the LUMO has mainly π-character. Both MOs exhibit A-symmetry in C_2. However, as the spectrum in Fig. 5.9 shows, excitation at $\lambda_p = 351$ nm does not lead to population of S_1. Instead a higher excited state – most probably S_2 – is populated. Transition to S_2 is preferentially governed by a LUMO+1$\leftarrow$HOMO excitation. The LUMO+1 is depicted in the right part of the red frame in Fig. 5.10. The latter mostly differs by symmetry (B) from the LUMO, which could account for the differences in both experimentally and theoretically observed oscillator strengths.

A very likely reason for the differences in excited states dynamics between Bz and TMB on the one hand, and Me on the other hand, becomes obvious at closer inspection of the transition energies to singlet and triplet states as collected in Tab. 5.4: Bz and TMB exhibit higher triplet states, which are nearly isoenergetic with the excited singlet states initially populated by light absorption. As discussed earlier, ISC efficiency is very sensitive to the energy gap between involved singlet and triplet states.[225] The calculated energy gap between S_1 and T_3 is 0.01 eV at the Franck-Condon geometry. Likewise, the energy gap between S_1 and T_4 of TMB is calculated to be 0.03 eV. Me, in contrast, exhibits an energy gap of 0.25 eV between the initially populated S_2 and the T_6 state, which is by at least a factor of 10 larger than in Bz and TMB. Also the S_1 state does not exhibit a triplet state nearly as close in energy as in the case of Bz and TMB.

The finding can also be correlated to the character of the involved singlet states. In Bz and TMB the two half occupied MOs are spatially well separated, which prevents the energy of the excited state from changing considerably with the change of one electron spin. An energetically close triplet state with the same MO occupation pattern can therefore be expected. In Me, on the other hand, spacial overlap and therewith influence on the energy of the state is by far more likely.

The overall errors in excitation energies from TDDFT calculations with the B3LYP functional can be estimated to be in the range of 0.2 eV (see Ref. 111). Thus, the values given in Tab. 5.4 are given definitively too precisely. However, the relative errors in excitation energy between singlet and triplet states, which are nearly isoenergetic due to their electronic structure, are probably considerably lower. Thus the accuracy of the values is legitimate in this case.

Moreover, it should be mentioned that a reduction of the discussion of the excited states properties to the Franck-Condon region is definitely a strong simplification. Nevertheless, the experimentally observed behavior is nicely recovered by this simplified discussion.

5.5. Comparison with efficiency relations

The relations of incorporation efficiency obtained by PLP-experiments[7] can now be qualitatively explained: Due to the nature of their respective singlet states excited at $\lambda_p = 351$ nm, the three photoinitiators exhibit largely differing excitation probabilities. The differences between Bz and TMB thereby can mainly be attributed to the shift of the spectra with respect to each other and the chemical nature of the produced benzyl and mesityl radicals, since the characters of both ground states and excited states are comparable. The singlet state which is initially populated by the excitation is S_1 in the case of Bz and TMB and S_2 in the case of Me. The nature of S_1 in Bz and TMB permits the existence of an isoenergetic excited triplet state, which allows for an efficient ISC. In Me no isoenergetic triplet state is available. Usually, excitation to a higher singlet state leads to occurrence of a growing number of additional depopulation channels. Since the density of states generally grows drastically with excitation energy, the probability of accessible conical intersections leading to other singlet states becomes higher. One of such depopulation channels, internal conversion into S_1, is predicted by Kasha's rule.[9] Therefore, only a minor part of population is brought into the triplet manifold in the case of Me. The observed time constants for the S_1 (S_2) depopulation in Bz, TMB, and Me therefore fit well to the results from the calculations.

Thus, the observations point to ISC as the crucial process dominating the efficiency of the photoinitiators. It can be estimated, if additional factors like the efficiency of α-cleavage and the initiation kinetics with the monomer play a significant role. For this purpose, the efficiency relations between excited photoinitiator molecules, which were yielded in Sect. 5.2, have to be compared to the findings from time-resolved spectroscopy. Excited TMB is between 1.8 and 2.2 times likelier to produce radicals, which are incorporated, than Bz. The probability of excited Bz to produce radicals, which are incorporated, is by a factor of 144 higher than for Me. It is now desirable to establish an efficiency relation between TMB and Me from the above data. In this case the relation would purely mirror excited states effects, since the initiating radical species, the mesitoyl radical, is the same (see Fig. 5.1).

As both the excited states structures and dynamics are comparable in Bz and TMB, the observed efficiency relation is attributed to differences in the ability of the produced benzyl and mesityl radicals to start polymerization. Mesityl radicals are therefore assumed to exhibit an incorporation probability, which is by an averaged factor of 2 higher than for benzyl radicals. Thus, the incorporation efficiency relation, which was obtained for Bz and Me can now be transformed to a relation between TMB and Me employing this factor. The transformation results in the relation that excited TMB has a probability to produce radicals, which are incorporated into polymers, by a factor of 72 higher than Me.

These new relations are now compared to the relative amplitudes originating from global fits to the TA traces, which are listed in Tab. 5.2. They refer to the TA at λ_{pr} = 500 nm. TA at λ_{pr} = 470 nm is not suitable, because small influences from ground state bleaching could interfere in the case of Me, also TA at λ_{pr} = 600 nm is inappropriate, since Me does not show TA there. The amplitudes A_1 and A_3 represent the number of molecules in the initially excited state and the number of radicals, respectively, each convoluted with the connected absorptivity per molecule. Here the assumption is made that both the respective initially populated excited states and the radical products of bond cleavage exhibit comparable absorption properties. The assumption is most justified in the case of TMB and Me, since the same mesityl starter radicals are produced. As mentioned earlier, the initial TA of Me is observed to exhibit an intensity by a factor of 4 lower than in TMB. As τ_1 is equal to the time resolution of the experiment, a convolution of the dynamics with the instrument response function is assumed, which leads to seemingly slower dynamics exhibiting lower TA intensity. Therefore, the relative amplitude A_1 of Me is multiplied by a factor of 4. The relations between amplitudes A_1 and A_3 of TMB and Me are, therefore, 0.85 and 0.01, respectively. With the two values a new relation between excited TMB and Me can be established. According to them, excited TMB is by a factor of 82 likelier to produce radicals, which are incorporated, than Me. The value is in reasonably good agreement with the value of 72 obtained from PLP-experiments and the extinction coefficients.

5.6. Conclusion and outlook

Despite the large error bars of the above estimation, it can be stated that ISC efficiency is the dominant effect on the ability of the investigated photoinitiators to start polymerization. Radical kinetics can be estimated to a reasonably good accuracy to have an effect, which is almost two orders of magnitudes smaller than ISC efficiency. Furthermore, the efficiency of ISC can be estimated to a sufficiently high accuracy by comparably less demanding TDDFT calculations at the Franck-Condon geometries of the photoinitiators.

It has to be stressed that the choice of the excitation wavelength has to be done carefully. It is rather likely that the high ISC efficiencies of Bz and TMB break down, when the molecules are excited into absorption bands of higher electronic states, because additional exit channels for the excited state population are provided.

To get more precise information about the efficiency of the radical generating process, a (theoretical) investigation of the respective absorption spectra of the radicals could be an interesting approach. With the knowledge of the pump pulse energy and the OD of the sample, the exact number of molecules, which are excited, can be calculated. If the absorption spectra of the radicals are known, λ_{pr} can be tuned to a spectral region, where the absorption can be exclusively attributed to a specific radical species. By obtaining TA intensities of two photoinitiators generating the same radical species (e.g. TMB and Me) and additionally varying the photoinitiator concentrations, absolute quantum yields should be achievable.

6. Excited state dynamics of photodepletable triplet photoinitiators

6.1. Introduction

The work presented in this chapter has been conducted in collaboration with Joachim Fischer in the group of Martin Wegener, Institute of Applied Physics, Karlsruhe Institute of Technology, and has in part been published.[226] The molecules investigated in the framework of this collaboration are – as the molecules discussed in the previous chapter – to be used as triplet photoinitiators. However, their special application leads to considerably different requirements with respect to their excited states properties.

The invention of stimulated emission depletion (STED) microscopy by Stephan W. Hell[241–245] gave biologists and medical scientists a unique tool to circumvent Abbe's diffraction limit of resolution in optical microscopy. Within classical fluorescence microscopy a sample, which is marked with a fluorescence dye, is scanned by a laser focused to the diffraction limit. Dye molecules are only excited, if they are within the illuminated area. Their fluorescence spectrum is red-shifted with respect to the wavelength of the excitation laser. Fluorescence can, therefore, be easily optically separated and detected. The idea of STED microscopy is to further reduce the effective area of the excitation laser spot by superimposing it with the torus-shaped mode of a depletion laser, which is tuned to stimulate emission from the dye molecules. Normal fluorescence spectra are broad and the spectrum of the depletion laser can be chosen to be considerably smaller. Thus, the fluorescence coming from the center of the area illuminated by the excitation laser has a different wavelength compared to the stimulated

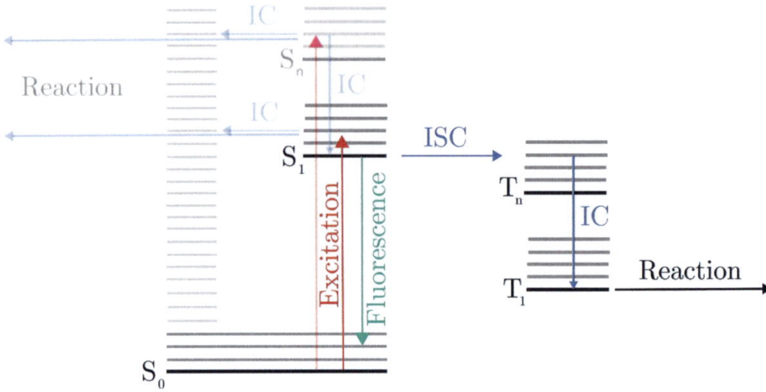

Figure 6.1.: Jabłoński-type diagram of the excited state processes following photoexcitation of a photoinitiator molecule, adapted from Fig. 1.1. The processes, which are relevant in this chapter are highlighted. They are relaxation via fluorescence into S_0 and ISC into the triplet manifold followed by radical formation from T_1.

emission coming from its outer parts. It can be therefore easily separated from the stimulated emission. Recently, the idea of STED was applied to the field of lithography.[8, 18, 92, 246] Here, instead of fluorescence dyes, triplet photoinitiators dissolved in a photoresin are excited by a laser to induce polymerization.

These photoinitiators are required to exhibit special properties which are best discussed with the Jabłoński-type diagram (see Fig. 6.1) already introduced in the prior chapters. In comparison to Fig. 1.1, the processes relevant for the present chapter are highlighted. The most obvious difference to convenient photoinitiators (see e.g. the photoinitiators investigated in Chap. 5) is that systems suitable for STED lithography have to exhibit considerable fluorescence quantum yields, to allow them to be depleted by stimulated emission. This requirement excludes at the same time measurable influence of ultrafast radiationless depopulation channels via IC. On the other hand, for photoinduced polymerization, an ISC channel to the triplet manifold has to be present and to be able to compete with

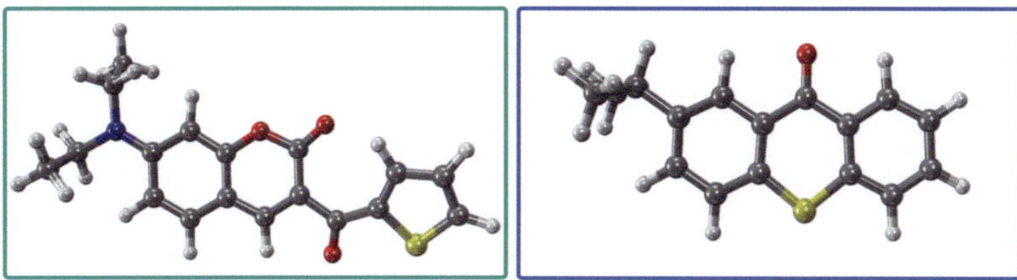

Figure 6.2.: Structures of DETC (green frame) and the 2-isopropyl-9-thioxanthenone isomer of ITX (blue frame) optimized with DFT.

fluorescence.

Two photoinitiator molecules, which meet these requirements, are 7-diethylamino-3-thenoylcoumarin (DETC) and isopropylthioxanthone (ITX), a mixture of the isomers 2-isopropyl-9-thioxanthenone and 4-isopropyl-9-thioxanthenone. Their structures are depicted in Fig. 6.2. Both molecules belong to the group of type II photoinitiators (see Chap. 5.1). DETC belongs to the group of ketocoumarin fluorescence dyes, which were also, in some cases, used as photoinitiators before.[221,247,248] Its fluorescence quantum efficiency is measured to be 2.5 % in ethanol soulution.[18] ITX belongs to the group of thioxanthones, which are widely used as photoinitiators.[209,223,227,249,250] Its fluorescence quantum yield is 15 % in ethanol solution.[92]

Although the ability of the excitation/depletion lithography setup to produce structures beyond the diffraction limit could be proved, the character of the depletion mechanism was not unambiguous, since different depletion mechanisms were proposed.[251,252] To clarify the underlying depletion mechanism, the excited states dynamics are investigated by femtosecond transient absorption (TA) spectroscopy. An additional goal of the investigation is to explain, why DETC yields much better results (i.e. much smaller structures in lithography) than ITX despite of its lower fluorescence quantum yield.

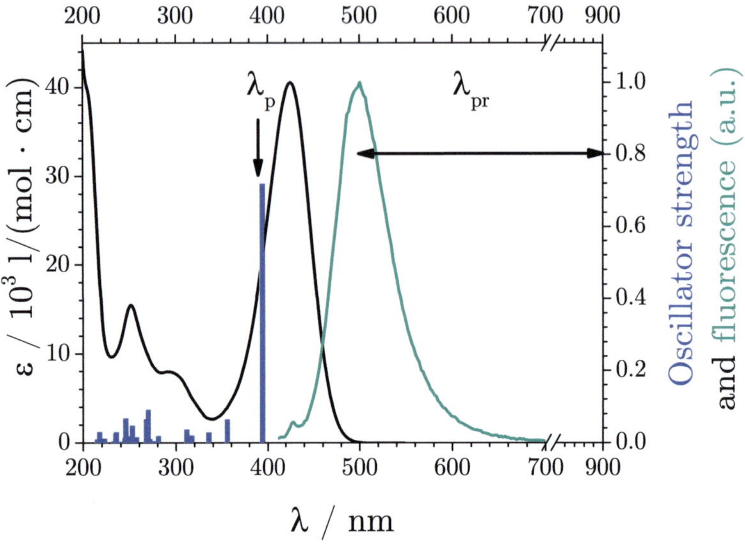

Figure 6.3.: Absorption (black) and fluorescence spectra (green) of DETC. Additionally, excitation energies (blue) calculated by TDDFT are inserted. Also λ_p and the λ_{pr} range are marked.

6.2. Absorption and fluorescence spectra of the photoinitiators

Absorption spectra of DETC and ITX in ethanol solution are depicted in Figs. 6.3 and 6.4, respectively, together with transitions from TDDFT calculations (for details see Chap. 3.6, geometries are listed in Tabs. C.12 and C.14 in appendix C) and fluorescence spectra. In the case of ITX the calculated transitions refer to the 2-isopropyl-9-thioxanthenone isomer. The transitions of the 4-isopropyl-9-thioxanthenone do not show large differences. A comparison between the calculated transitions of both isomers is given in the appendix B.1. The agreement between experimental absorption spectrum and theoretical calculated values is reasonably good. It has to be taken into account that no solvent effect was included into the calculations.

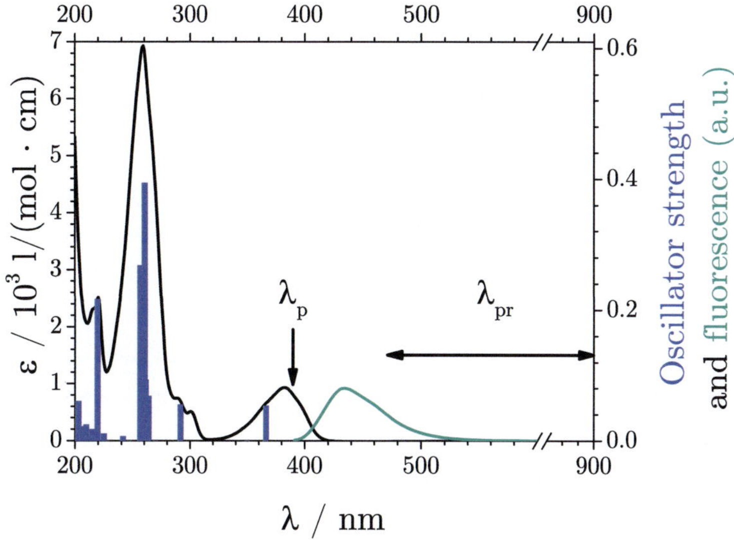

Figure 6.4.: Absorption (black) and fluorescence spectra (green) of ITX. Additionally, excitation energies (blue) calculated by TDDFT are inserted. Also λ_p and the λ_{pr} range are marked.

Since solvation in most cases leads to a red-shift of the spectrum, part of the discrepancy can be attributed to this effect. The respective first absorption bands can be, in both cases, unambiguously assigned to excitation into the first excited singlet state S_1. Accordingly, excitation at $\lambda_p = 388$ nm exclusively brings both molecules into the first excited singlet state (note the arrows in Figs. 6.3 and 6.4 labeling the position of λ_p).

Comparing Figs. 6.3 and 6.4, both calculations and experiment show the same differing behavior in the global absorption maxima: In the spectrum of DETC the first absorption band centered at $\lambda = 425$ nm is the by far strongest (the high extinction coefficients at $\lambda \approx 200$ nm are partly due to solvent absorption[253]). In the spectrum of ITX the first absorption band at $\lambda = 383$ nm is by far not the strongest. The strongest is instead centered at $\lambda = 259$ nm. Furthermore, the values of the extinction coefficients differ substantially. In the DETC spectrum the first absorption band exhibits a maximum extinction coefficient of $\varepsilon = 4.06 \cdot 10^4$ l/(mol $\cdot$ cm), the first absorption band of ITX exhibits only $\varepsilon = 9.32 \cdot 10^2$ l/(mol $\cdot$ cm). Also the extinction coefficient of the strongest absorption band of ITX ($\varepsilon = 6.93 \cdot 10^3$ l/(mol $\cdot$ cm)) is considerably lower than the first absorption band of DETC. The DETC spectrum resembles the spectra of typical fluorescence dyes (see e.g. the spectrum of Stilbene 3[254]) both in shape and in the range of extinction coefficient values. With its weak first absorption band the ITX spectrum rather reminds of the photoinitiator spectra presented in Chap. 5. As the oscillator strength for stimulated emission – like for absorption – is strongly dependent on the coupling between ground state and first excited singlet state, the radiative relaxation properties of DETC are expected to be better than for ITX. However, the fluorescence quantum yields of 2.5 % (DETC) and 15 % (ITX) seem to contradict this expectation.

6.3. Time-resolved investigation of the excited states dynamics

TA traces of DETC at different λ_{pr} are depicted in Fig. 6.5. It can be nicely seen that at short λ_{pr} the TA is purely negative, and at long λ_{pr} purely positive. At intermediate wavelengths a superposition of both borderline cases can be observed. As discussed in Chap. 2.1.5, possible contributions to TA intensity are ground state bleaching, SEM, ESA, and absorption of the vibrationally "hot" ground state. The first two contributions yield negative TA values, the latter two positive values. Contribution of "hot" ground state absorption necessitates the existence of an ultrafast radiationless channel from S_1 into S_0. As both fluorescence and ISC are reported for DETC, a radiationless channel cannot be ruled out, but plays a minor role, if existing at all. Therefore, "hot" ground state absorption is not further taken into account.

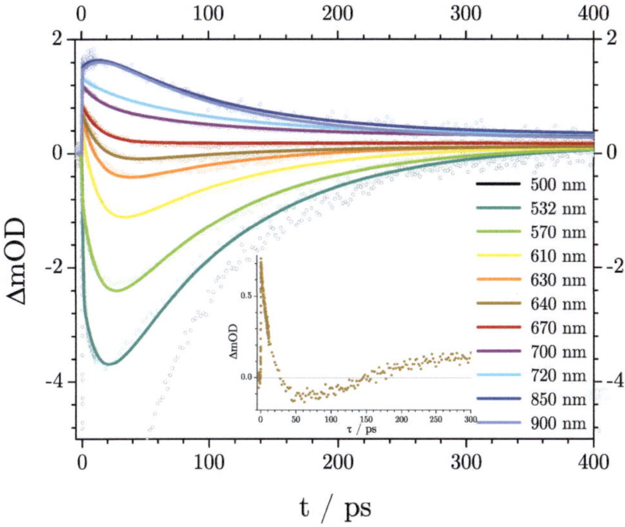

Figure 6.5.: TA traces (dots) and global fit analysis (lines) of DETC in ethanol solution at λ_p = 388 nm and several λ_{pr} between 500 nm and 900 nm. For the data of λ_{pr} = 500 nm no fit analysis is made, since it shows a contribution of ground state bleaching, which is not included in the fit. A more detailed plot of the TA trace at λ_{pr} = 640 nm is inserted.

Since λ_{pr} = 500 nm is at the red edge of the DETC absorption spectrum (see Fig. 6.3), in this case a considerable contribution of ground state bleaching cannot be excluded. At the remaining λ_{pr}, however, only SEM and ESA can contribute to TA intensity. At all λ_{pr}, where negative TA is observable, the TA minimum is not reached immediately after excitation, but within several tens of ps. This observation fits together well with the implications of Kasha's rule.[9] The constraint that fluorescence takes place from the S_1 minimum results in the need for a vibrational relaxation step, which brings population out of the Franck-Condon region of the S_1 PES into the S_1 minimum where fluorescence occurs. The dissipation of absorbed energy into vibrational modes is obvious by the red-shift of fluorescence spectra with respect to the excitation wavelength. Since in the TA experiments emission is also stimulated by λ_{pr} red-shifted to λ_p, this vibrational relaxation step can be observed in the TA data.

It has to be stressed that negative TA values must not be misinterpreted as an absence of ESA. In the spectral area of the DETC fluorescence (see Fig. 6.3) the sign of TA values is always only an indication which of both processes dominates. This is proved by comparing the respective TA minima with the shape of the DETC fluorescence band. They follow nicely the band shape and one could argue that ESA is a minor effect in this case. However, on closer inspection of the TA trace at $\lambda_{pr} = 640$ nm (see insertion in Fig. 6.5) one can observe a shallow TA minimum which hardly exhibits negative TA values. Since the fluorescence spectrum exhibits considerable intensity at 640 nm (see Fig. 6.3), the signal would have a by far stronger negative component, if no ESA was observable. However, in this case ESA lifts the signal into positive TA values and only its shape indicates the existence of SEM. The same considerations also hold for the TA at $\lambda_{pr} = 670$ nm.

Common fluorescence dyes exhibit fluorescence lifetimes in the range of ns. The decay of TA intensity on the time scale of 200 ps, which is obviously connected to population decay of S_1, therefore cannot be solely due to radiative relaxation. The competing relaxation channel is obvious, since DETC is known to be a triplet photoinitiator. A considerable amount of S_1 population thus has to be transferred to the triplet manifold by ISC on this time scale, followed by rapid relaxation into T_1. As stimulated emission from T_1 into S_0 is forbidden by the spin selection rule, at later delay times only positive TA originating from ESA of T_1 into higher triplet states is found. The TA pattern at long delay times (around 400 ps) shows that T_1 does not seem to exhibit large absorption bands in the visible.

To quantitatively evaluate the experimental findings, the above considerations on possible relaxation channels are summarized into the relaxation scheme presented in Fig. 6.6: The initial population in the Franck-Condon

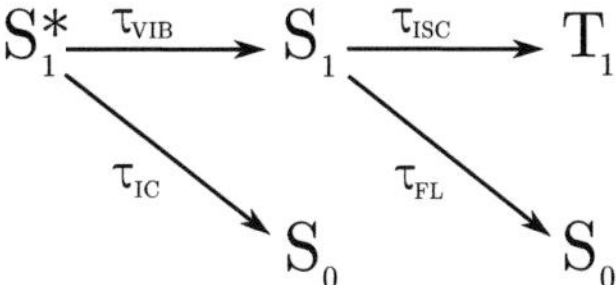

Figure 6.6.: Simple rate equation model as basis for generation of a fit function (see Appendix B.2): The vibrationally excited level (S_1^*) of the first excited singlet state is depopulated by two competing processes, vibrational relaxation into S_1 (time constant τ_{VIB}) and radiationless internal conversion into the ground state (S_0) (time constant τ_{IC}). S_1 is depopulated by competing channels leading to the lowest triplet state T_1 by intersystem crossing (τ_{ISC}) and to S_0 by fluorescence (τ_{FL})

region (S_1^*), which is prepared by the pump pulse, mainly relaxes into the S_1 minimum (labeled as S_1). As discussed above, a radiationless relaxation channel to S_0 cannot be ruled out. S_1 is depopulated via two competing channels, fluorescence and ISC. Population decay in T_1 cannot be observed, because it takes place outside of the investigated time window.

For evaluation of the data with a fit function, all relaxation processes are assumed to be statistical and can be therefore described by singly exponential functions depending on time constants τ_i. This assumption is potentially wrong in the case of radiationless relaxation into S_0. Since the channel is presumably very weak, the error is negligible. Hence, from the relaxation scheme a system of coupled differential equations is constituted, which can be solved analytically. The derivation is pointed out in detail in appendix B.2. It results in the function

$$\Delta OD(\tau) = Ae^{-\frac{\tau}{\tau_1}} + Be^{-\frac{\tau}{\tau_2}} + C\left[\tau_1 e^{-\frac{\tau}{\tau_1}} - \tau_2 e^{-\frac{\tau}{\tau_2}} + \tau_2 - \tau_1\right] \qquad (6.1)$$

with amplitudes A, B, and C depending on the extinction coefficients ε of the excited states S_1^*, S_1, and T_1 according to $(A+B) \propto \varepsilon_{S_1^*}$, $B \propto \varepsilon_{S_1}$, and $C \cdot (\tau_2 - \tau_1) \propto \varepsilon_{T_1}$. The time constants τ_1 and τ_2 depend on the time constants presented in Fig. 6.6 according to $\tau_1 = \frac{\tau_{VIB}\tau_{IC}}{\tau_{VIB}+\tau_{IC}}$ and $\tau_2 = \frac{\tau_{ISC}\tau_{FL}}{\tau_{ISC}+\tau_{FL}}$. τ_1 and τ_2, therefore, are the depopulation time constants of S_1^* and S_1,

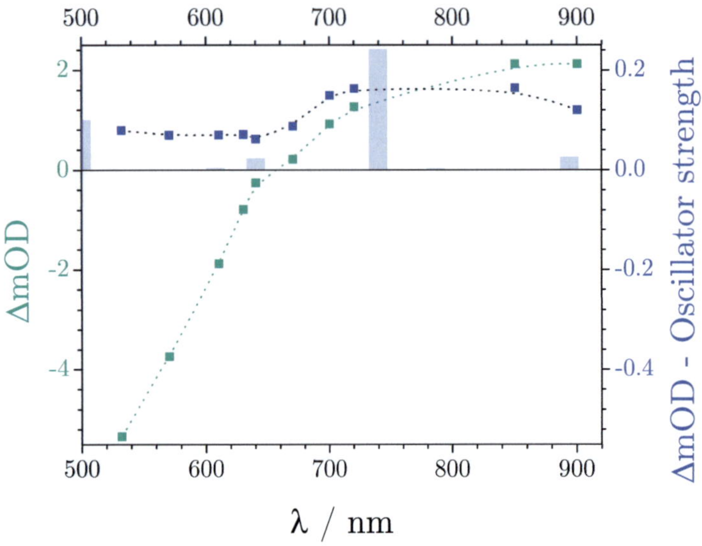

Figure 6.7.: Wavelength-dependent fit coefficients B (green) and C (blue) resulting from a global fit of the TA traces of DETC according to Eq. 6.1. Additionally, excitations (bars) calculated at a relaxed T_1 geometry with TDDFT are inserted.

respectively. The TA data of DETC are fitted by Eq. 6.1 in the way that time constants are optimized globally and amplitudes specifically for every λ_{pr}. The fit yields values of $\tau_1 = 13.8 \pm 0.2$ ps and $\tau_2 = 99 \pm 1$ ps. The time constants are also listed in Tab. 6.1. As in the preceding chapters, the errors for the time constants either result from the standard deviations yielded from the fitting routine or from the time resolution of the experiment. The

sample	τ_1 / ps	τ_2 / ns
DETC	13.8 ± 0.2	$(9.9 \pm 0.1) \cdot 10^{-2}$
ITX	48.0 ± 0.2	2.3 ± 0.2

Table 6.1.: Time constants resulting from global fits of the TA traces in Figs. 6.5 and 6.8 with Eq. 6.1.

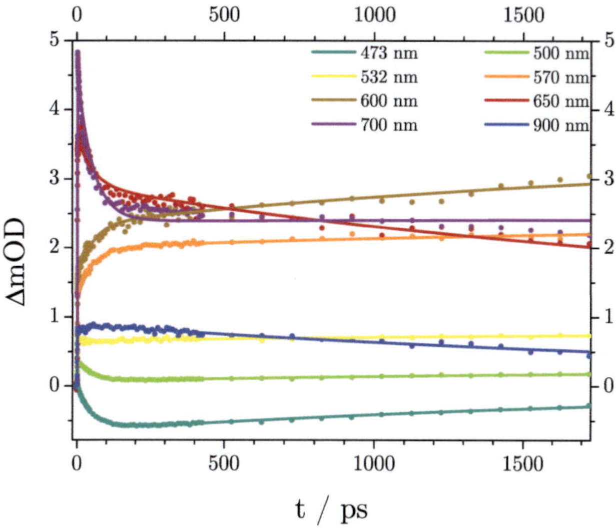

Figure 6.8.: TA traces (dots) and global fit analysis (lines) of ITX in ethanol solution at λ_p = 388 nm and several λ_{pr} between 473 nm and 900 nm.

TA at λ_{pr} = 500 nm was not included into the global fit, since contribution of ground state bleaching cannot be ruled out in this case.

The wavelength dependence of the amplitudes B and C is shown in Fig. 6.7. As could be expected from the TA traces, ε_{S_1}, which is proportional to B, shows a strong wavelength dependence, which is comparable to the TA behavior, with negative values at short wavelengths and positive values at long wavelengths. ε_{T_1}, which is proportional to $C \cdot (\tau_2 - \tau_1)$ exhibits, as expected, only positive values, which show a maximum in the long wavelength regime. This finding fits together well with triplet-triplet excitations calculated with TDDFT at a geometry optimized with respect to T_1 (see Tab. C.13 in appendix C for coordinates). They are inserted in Fig. 6.7.

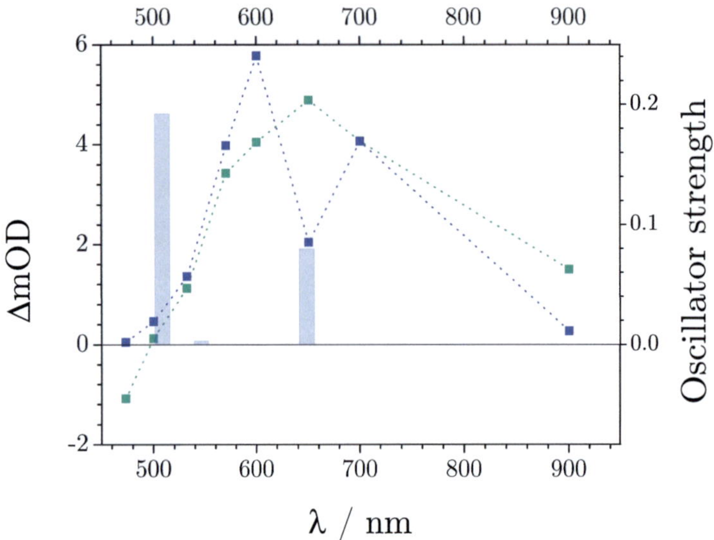

Figure 6.9.: Wavelength-dependent fit coefficients B (green) and C (blue) resulting from a global fit of the TA traces of ITX according to Eq. 6.1. Additionally, excitations (bars) calculated at a relaxed T_1 geometry with TDDFT are inserted.

The TA traces of ITX are depicted in Fig. 6.8. The dependence of the sign of the TA values on λ_{pr} is qualitatively similar to DETC. However, negative TA values are only found at the shortest employed λ_{pr} of 473 nm. The remaining traces exhibit only positive TA values. Since the absorption and fluorescence bands of ITX are considerably blue-shifted with respect to DETC (see Fig. 6.4), both ground state bleaching and "hot" ground state absorption can be neglected as contributions to the observed TA. The general shape of the TA traces is also quite comparable to DETC: At short λ_{pr} they exhibit minima, which cannot be observed at long λ_{pr}. The same reason for the existence of the minima holds as in the case of DETC: An initial vibrational relaxation step from the Franck-Condon region into the S_1 minimum, where radiative relaxation can take place.

The largest and most obvious difference compared to the TA of DETC is the, by far, slower time scale of relaxation. The observable dynamics of DETC end 400 ps after excitation, in the case of ITX the dynamics are obviously not completed on the experimentally accessible time scale of 1.6 ns. The finding is also revealed by the time constants $\tau_1 = 48 \pm 0.2$ ps and $\tau_2 = 2.3 \pm 0.2$ ns resulting from a global fit with the same function as for DETC (Eq. 6.1 and Tab. 6.1). The fit coefficients proportional to ε_{S_1} and ε_{T_1} are depicted in Fig. 6.9 together with calculated triplet-triplet excitations at a geometry optimized with respect to T_1 (see Tab. C.16 in appendix C for coordinates). As opposed to DETC the values proportional to ε_{S_1} are, even at the shortest λ_{pr}, hardly negative. The value of τ_2 is in surprisingly good agreement with earlier observed fluorescence lifetimes.[92]

6.4. Discussion

The comparison between DETC and ITX can be summed up in the way that the S_1 state of DETC exhibits a strong coupling to both S_0 and the triplet manifold. The coupling properties to S_0 are visible on the one hand in the relative and absolute values of the extinction coefficients of the first absorption band, which can be connected to S_1. On the other hand, the strongly negative part of both TA and the evaluated fit coefficients point in this direction. The coupling to the triplet manifold can be estimated from the value of τ_2, which is, at least, one order of magnitude below typical S_1 lifetimes of fluorescence dyes.

The S_1 state of ITX, in contrast, shows neither a strong coupling to S_0 nor to the triplet manifold. The former is observable by the low extinction coefficients of the first absorption band in the static spectrum and by the mostly positive values of TA and the fit coefficients. The latter is apparent by the comparably high value of τ_2.

Based on the properties discussed so far, ITX seems to be the more appropriate photoinitiator for STED lithography, because its S_1 lifetime is considerably higher than in DETC. Therefore time dependence is a minor problem and depletion can be done with a cw laser. However, as lithography experiments reveal, even with the employment of a cw laser for SEM depletion, DETC gives considerably better results than ITX.[8, 18, 91, 246, 255] The reason lies in the superposition of ESA and SEM of S_1 at the wavelength of the employed depletion laser. For this purpose a convenient frequency doubled Nd:YAG laser (532 nm) was employed. At 532 nm DETC exhibits a strongly negative ε_{S_1}, and a comparably low positive ε_{T_1}. ITX in contrast has a slightly positive ε_{S_1}, and also a low positive ε_{T_1}. Again the positive ε_{S_1} does not mean that SEM is not taking place at all. However, the probability of ESA exceeds the probability of SEM. The effect of ESA on the system is that additional energy is brought into the sample by absorption of a depletion photon in contrast to the effect of SEM, where energy is extracted from the system by a leaving photon. The additional amount of energy introduced by ESA brings the photoinitiator molecule into higher excited states. There, the efficiency of ISC can be enlarged and additional nonradiative channels to S_0 can occur. Both effects lower the depletion efficiency.

6.5. Conclusion and outlook

The results from static and transient absorption spectroscopy as well as from quantum chemical calculations clearly point to a stimulated emission depletion mechanism being responsible for the achievements with DETC in STED lithography. However, this is not as clear in the case of ITX, since the probability for ESA is higher than for SEM at the depletion wavelength of 532 nm employed in the lithography experiments. Thus, for ITX different mechanisms like the ones proposed earlier[251, 252] also have to be taken into account.

Furthermore, it can be summarized that the fluorescence quantum yield is not an appropriate measure for the suitability of a photoinitiator for STED lithography. Additional factors, namely the S_1 lifetime and the relation between the probabilities of ESA and SE, are equally critical.

The mechanism of SEM depletion in DETC is meanwhile additionally supported by new lithography experiments.[18] In these experiments a pulsed depletion source with a tunable time delay to the excitation laser pulse was employed. Two depletion components could be observed, whereas the dependence of the stronger component on the time delay and the fluorescence spectrum is in good agreement with the SEM mechanism. However, an additional, weaker component was observed, which shows no time dependence.[8,18] Further attempts are now aimed at improving the efficiency of the lithography process by systematic design of new photoinitiator species based on DETC.

A. Appendix for Chapter 4

A.1. Synthesis of C_5Br_6

The synthesis of C_5Br_6 by bromination of C_5H_6 under strongly basic conditions proceeds most likely via deprotonation of the sp^3 hybridized carbon ($pK_a = 18$, see Ref. 256), which leads to a stable, highly symmetric aromatic anion (D_{5h}). In a second step the anion is attacked by hypobromic acid leading to a monobrominated cyclopentadiene. This species can be deprotonated again etc. Since the fully brominated compound is easily ob-

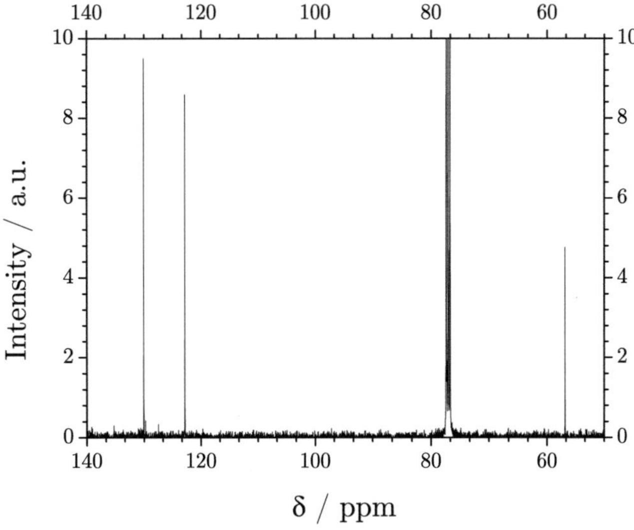

Figure A.1.: ^{13}C-NMR spectrum of the synthesized C_5Br_6. The peaks at 77 ppm result from the solvent CDCl$_3$.

tained with a high yield, the attack of hypobromic acid must regioselectively lead to bromination at a ring position, which is yet substituted by a hydrogen.

The procedure for the synthesis of C_5Br_6 is adapted from Ref. 83. All chemicals where obtained from Sigma Aldrich and used without further purification. Dicyclopentadiene was depolymerized via distillation into an ice-cooled flask. The distillation product was directly processed. 144 g of Potassium hydroxide were dissolved in 450 ml of water while cooling in ice. Within 1 h, 20 ml of bromine were added in droplets under stirring and cooling in ice. After complete addition of bromine, cooling was stopped and a solution of 4.5 ml of cyclopentadiene in 30 ml hexane was added in droplets within a few minutes while stirring. The mixture was stirred over night and then filtered. The aqueous and organic phases were separated. The solid residue from filtering and the organic phase were extracted several times with hexane. The combined extracts were concentrated until a dark brown solid precipitated from the solution at -18 °C. The solid was filtered out, recrystallized in pentane and dried in vacuo. It was characterized by ^{13}C-NMR (see Fig. A.1 and Ref. 257) and UV-Vis spectroscopy. By comparing peak integrals of C_5Br_6 with the ones of side products, the purity of the product is estimated to be >95 %.

A.2. Decay associated spectra (DAS) resulting from fits of the TRPE spectra

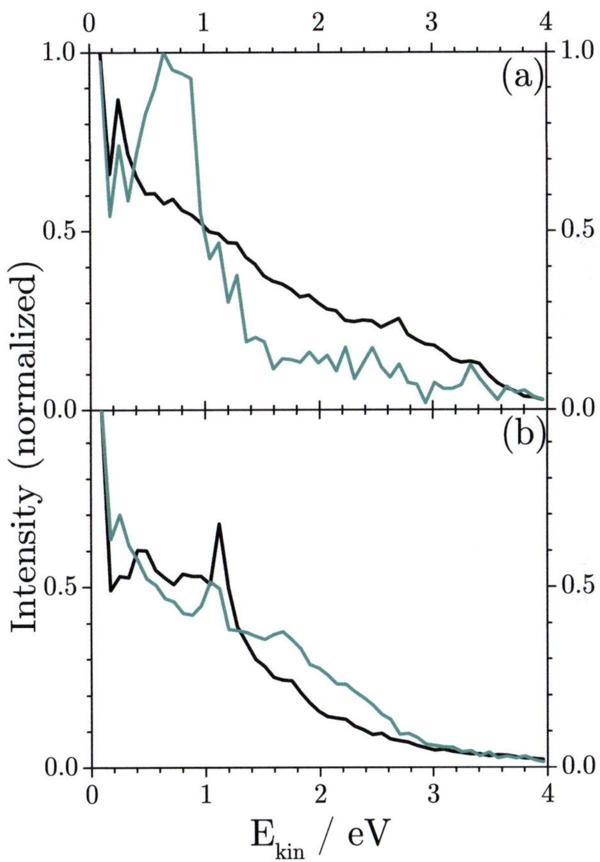

Figure A.2.: DAS associated with τ_1 (black) and τ_2 (green) resulting from fits (a) of the TRPE spectrum of C_5Cl_6 in Fig. 4.9 (a) at λ_{pr} = 267 nm and (b) of the TRPE spectrum in Fig. 4.9 (b) at λ_{pr} = 400 nm with functions of the type of Eq. 4.2.

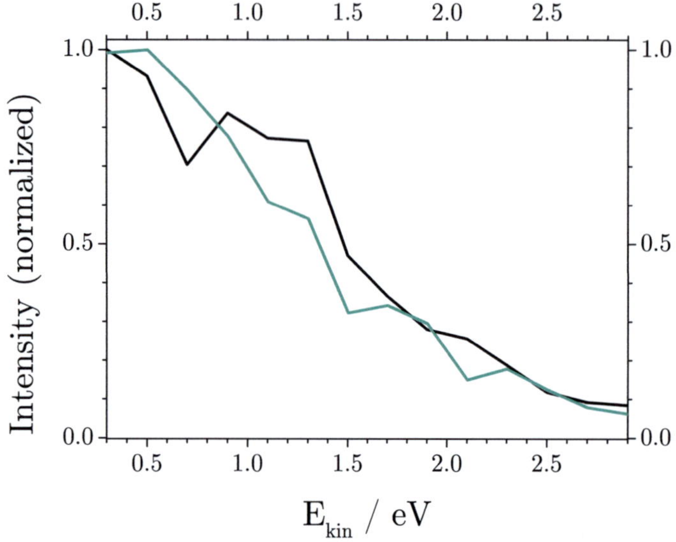

Figure A.3.: DAS associated with τ_1 (black) and τ_2 (green) resulting from fits of the TRPE spectrum of C_5Br_6 in Fig. 4.10 with functions of the type of Eq. 4.2.

B. Appendix for Chapter 6

B.1. Comparison of the calculated electronic transitions of the two ITX isomers

The optimized geometries of the two isomers of ITX, 2-isopropyl-9-thio-xanthenone and 4-isopropyl-9-thioxanthenone are listed in Tabs. C.14 and C.15. Comparisons of excitations with A' and A" symmetry, calculated with TDDFT, are shown in Fig. B.1 (a) and (b). As can be seen in the two figures, the differences are irrelevant at least for the lowest excited states, which were investigated experimentally.

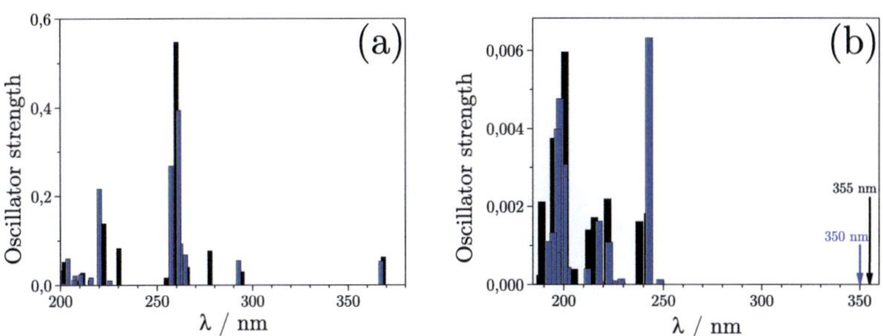

Figure B.1.: Comparison of electronic transitions with (a) A' and (b) A" symmetry of 2-isopropyl-9-thioxanthenone (black) and 4-isopropyl-9-thio-
xanthenone (blue) calculated with TDDFT.

B.2. Derivation of the fit function

Kinetic analysis of the relaxation scheme depicted in Fig. 6.6 yields three coupled differential equations describing the population of S_1^*, S_1 and T_1:

$$\frac{dS_1^*}{dt} = -\frac{1}{\tau_{\text{IC}}} S_1^* - \frac{1}{\tau_{\text{VIB}}} S_1^* = -\underbrace{\frac{\tau_{\text{VIB}} + \tau_{\text{IC}}}{\tau_{\text{VIB}} \tau_{\text{IC}}}}_{\tau_1^{-1}} S_1^* \tag{B.1}$$

$$\frac{dS_1}{dt} = \frac{1}{\tau_{\text{VIB}}} S_1^* - \frac{1}{\tau_{\text{ISC}}} S_1 - \frac{1}{\tau_{\text{FL}}} S_1 = \frac{1}{\tau_{\text{VIB}}} S_1^* - \underbrace{\frac{\tau_{\text{ISC}} + \tau_{\text{FL}}}{\tau_{\text{ISC}} \tau_{\text{FL}}}}_{\tau_2^{-1}} S_1 \tag{B.2}$$

$$\frac{dT_1}{dt} = \frac{1}{\tau_{\text{ISC}}} S_1 \tag{B.3}$$

Population of S_0 can be neglected, because the contribution of vibrationally "hot" ground state absorption can be estimated as very small.

The differential equations B.1 - B.3 can now be solved employing the boundary condition that the initial population of state S_1^* is $S_1^*(t=0) = S_{1,0}^*$ and 0 for the other two states. Integration of equation B.1 yields:

$$S_1^*(t) = S_{1,0}^* \cdot e^{-\frac{t}{\tau_1}} \tag{B.4}$$

Equation B.2 is an inhomogeneous differential equation and can be solved by inserting Equation B.4, written as

$$\frac{dS_1}{dt} + \frac{1}{\tau_2} S_1 = \frac{1}{\tau_{\text{VIB}}} S_{1,0}^* e^{-\frac{t}{\tau_1}} . \tag{B.5}$$

The homogeneous differential equation

$$\frac{dS_1}{dt} + \frac{1}{\tau_2} S_1 = 0 \tag{B.6}$$

can be integrated yielding a general solution

$$S_1 = e^{-\frac{t}{\tau_2}} \cdot c. \tag{B.7}$$

The particular solution is generated by inserting the general solution into equation B.5 and varying the constant c:

$$\frac{d}{dt}\left[e^{-\frac{t}{\tau_2}} \cdot c(t)\right] + \frac{1}{\tau_2}e^{-\frac{t}{\tau_2}} \cdot c(t) = \frac{1}{\tau_{\mathrm{VIB}}}S_{1,0}^* e^{-\frac{t}{\tau_1}} \tag{B.8}$$

This equation has to be solved for c(t):

$$\frac{dc}{dt} = \frac{1}{\tau_{\mathrm{VIB}}}S_{1,0}^* e^{-t\left(\frac{1}{\tau_1}-\frac{1}{\tau_2}\right)} \tag{B.9}$$

$$c(t) = -\frac{\tau_1 \cdot \tau_2}{\tau_{\mathrm{VIB}}(\tau_2 - \tau_1)}S_{1,0}^* e^{-t\left(\frac{\tau_2-\tau_1}{\tau_2\cdot\tau_1}\right)} + f \tag{B.10}$$

The sum of equations B.10 and B.7

$$S_1(t) = e^{-\frac{t}{\tau_2}}\left[c+f - \frac{\tau_1 \cdot \tau_2}{\tau_{\mathrm{VIB}}(\tau_2 - \tau_1)}S_{1,0}^* e^{-t\left(\frac{\tau_2-\tau_1}{\tau_2\cdot\tau_1}\right)}\right] \tag{B.11}$$

is the solution for differential equation B.2. Inserting the boundary condition for S_1 yields:

$$S_1(t) = S_{1,0}^* \underbrace{\frac{\tau_1 \cdot \tau_2}{\tau_{\mathrm{VIB}}(\tau_2 - \tau_1)}}_{G} e^{-\frac{t}{\tau_2}}\left[1 - e^{-t\left(\frac{\tau_2-\tau_1}{\tau_2\cdot\tau_1}\right)}\right] \tag{B.12}$$

Integration of differential equation B.3 gives

$$T_1(t) = S_{1,0}^* \frac{\tau_1 \cdot \tau_2}{\tau_{\mathrm{ISC}}\,\tau_{\mathrm{VIB}}(\tau_2 - \tau_1)}\left[\tau_1 e^{-\frac{t}{\tau_1}} - \tau_2 e^{-\frac{t}{\tau_2}}\right] + c \tag{B.13}$$

and with the boundary condition for T_1

$$T_1 = S_{1,0}^* \frac{G}{\tau_{\mathrm{ISC}}} \left[\tau_1 e^{-\frac{t}{\tau_1}} - \tau_2 e^{-\frac{t}{\tau_2}} + \tau_2 - \tau_1 \right]. \tag{B.14}$$

The contributions of states S_1^*, S_1, and T_1 to ΔOD are:

$$\Delta OD_{S_1^*}(t) = \varepsilon_{S_1^*} S_1^*(t) = \varepsilon_{S_1^*} S_{1,0}^* e^{-\frac{t}{\tau_1}}, \tag{B.15}$$

$$\Delta OD_{S_1}(t) = \varepsilon_{S_1} S_{1,0}^* G \left[1 - e^{-t\left(\frac{\tau_2 - \tau_1}{\tau_2 \cdot \tau_1}\right)} \right], \tag{B.16}$$

and

$$\Delta OD_{T_1}(t) = \varepsilon_{T_1} S_{1,0}^* \frac{G}{\tau_{\mathrm{ISC}}} \left[\tau_1 e^{-\frac{t}{\tau_1}} - \tau_2 e^{-\frac{t}{\tau_2}} + \tau_2 - \tau_1 \right], \tag{B.17}$$

where $\varepsilon_{S_1^*}$, ε_{S_1}, and ε_{T_1} are extinction coefficients of the respective states at λ_{pr}. The total ΔOD is given by

$$\Delta OD_{\mathrm{tot}}(\tau) = g(\tau) \left(e^{-\frac{\tau}{\tau_1}} \underbrace{S_{1,0}^* \left[\varepsilon_{S_1^*} - \varepsilon_{S_1} G \right]}_{A} + e^{-\frac{\tau}{\tau_2}} \underbrace{S_{1,0}^* \varepsilon_{S_1} G}_{B} \right.$$

$$\left. + \left[\tau_1 e^{-\frac{\tau}{\tau_1}} - \tau_2 e^{-\frac{\tau}{\tau_2}} + \tau_2 - \tau_1 \right] \underbrace{S_{1,0}^* \varepsilon_{T_1} \frac{G}{\tau_{\mathrm{ISC}}}}_{C} \right), \tag{B.18}$$

where the coefficients (A+B), B, and C are each dependent on a differing constant factor, which consists of the initial population $S_{1,0}^*$, an expression of time constants, and only one of the extinction coefficients $\varepsilon_{S_1^*}$, ε_{S_1}, and ε_{T_1}. The time variable t is now replaced by the time delay τ between pump and probe pulses and the expression is multiplied with the instrument response function $g(\tau)$ (see Eq. 3.3).

C. Geometry data of calculated species

C.1. C_5Cl_6

Table C.1.: Minimum geometry of C_5Cl_6

Atom	x	y	z
C	0.00000	0.74392	0.00000
C	0.00000	-0.74392	0.00000
C	0.00000	1.19602	1.28459
C	0.00000	-1.19602	1.28459
C	0.00000	0.00000	2.22626
Cl	0.00000	1.68047	-1.43343
Cl	0.00000	-1.68047	-1.43343
Cl	0.00000	2.80719	1.84983
Cl	0.00000	-2.80719	1.84983
Cl	1.48322	0.00000	3.25346
Cl	-1.48322	0.00000	3.25346

Table C.1.: Geometry data of the ground state minimum of C_5Cl_6 optimized by DFT. The geometry is at the same time the start geometry of the interpolated path between C_5Cl_6 and the CT complex (see Fig. 4.13). It is given in Cartesian coordinates and units of angstroms.

Table C.2.: Minimum geometry of $C_5Cl_5\cdots Cl$

Atom	x	y	z
C	0.00000	0.74454	0.00000
C	0.00000	-0.74454	0.00000
C	0.00000	1.18879	1.3073
C	0.00000	-1.18879	1.3073
C	-0.04733	0.00000	2.14787
Cl	-0.03878	1.68945	-1.41784
Cl	-0.03878	-1.68945	-1.41784
Cl	-0.00344	2.80337	1.88111
Cl	-0.00344	-2.80337	1.88111
Cl	0.04455	0.00000	3.79949
Cl	2.24138	0.00000	5.00388

Table C.2.: Geometry data of the ground state minimum of the $C_5Cl_5\cdots Cl$ CT complex optimized by DFT. The geometry is at the same time the end geometry of the interpolated path between C_5Cl_6 and the CT complex (see Fig. 4.13). It is given in Cartesian coordinates and units of angstroms.

Table C.3.: Minimum geometry of C_5Cl_5

Atom	x	y	z
C	-1.163062	-0.000591	-0.362146
C	-0.756834	-0.000441	0.940437
C	0.738649	0.000917	0.954805
C	1.169973	0.000657	-0.339784
C	0.011431	0.000534	-1.181706
Cl	-2.772244	-0.000344	-0.941905
Cl	0.027665	-0.000444	-2.878775
Cl	2.789492	0.000771	-0.889810
Cl	1.675695	-0.001333	2.369513
Cl	-1.720660	0.000985	2.337044

Table C.3.: Geometry data of the ground state minimum of C_5Cl_5 optimized by DFT. It is given in Cartesian coordinates and units of angstroms.

C.2. C_5Br_6

Table C.4.: Minimum geometry of C_5Br_6

Atom	x	y	z
C	0.00000	0.00000	0.74184
C	0.00000	0.00000	-0.74184
C	1.27662	0.00000	1.18993
C	1.27662	0.00000	-1.18993
C	2.21424	0.00000	0.00000
Br	-1.55234	0.00000	1.78139
Br	-1.55234	0.00000	-1.78139
Br	1.8708	0.00000	2.95723
Br	1.8708	0.00000	-2.95723
Br	3.33524	1.61884	0.00000
Br	3.33524	-1.61884	0.00000

Table C.4.: Geometry data of the ground state minimum of C_5Br_6 optimized by DFT. The geometry is at the same time the start geometry of the interpolated path between C_5Br_6 and the CT complex (see Fig. 4.15). It is given in Cartesian coordinates and units of angstroms.

Table C.5.: Minimum geometry of $C_5Br_5\cdots Br$

Atom	x	y	z
C	0.00000	0.00000	0.73946
C	0.00000	0.00000	-0.73946
C	1.27147	0.00000	1.17467
C	1.27147	0.00000	-1.17467
C	2.13179	-0.04569	0.00000
Br	-1.53542	-0.02466	1.78139
Br	-1.53542	-0.02466	-1.78139
Br	1.88046	3.4E-4	2.93313
Br	1.88046	3.4E-4	-2.93313
Br	4.89095	2.62874	0.00000
Br	3.86937	0.12956	0.00000

Table C.5.: Geometry data of the ground state minimum of the $C_5Br_5\cdots Br$ CT complex optimized by DFT. The geometry is at the same time the end geometry of the interpolated path between C_5Br_6 and the CT complex (see Fig. 4.15). It is given in Cartesian coordinates and units of angstroms.

C.3. Benzoin (Bz)

Table C.6.: Bz Geometry 1

Atom	x	y	z
C	-2.9097956	1.3252262	-0.1264521
C	-1.9679171	1.0851780	0.8809709
C	-2.3327736	1.3027020	2.2175136
C	-3.6114225	1.7602067	2.5359998
C	-4.5441504	2.0039555	1.5226982
C	-4.1920484	1.7820727	0.1898696
C	-0.5870494	0.5379721	0.5356083
O	-0.4577053	-0.8093478	0.9487938
C	0.5428011	1.3629320	1.1958928
O	1.2165488	0.7989638	2.0584039
C	0.8129148	2.7705224	0.8000565
C	0.1189439	3.4143401	-0.2393503
C	0.4272846	4.7306709	-0.5820239
C	1.4292811	5.4181599	0.1085574
C	2.1265707	4.7853583	1.1443014
C	1.8224839	3.4711185	1.4863371
H	2.9089532	5.3213027	1.6831659
H	-0.1169790	5.2227959	-1.3888542
H	0.2540749	-0.7822068	1.6327314
H	-2.6414884	1.1365118	-1.1690847
H	-4.9187696	1.9558925	-0.6053045
H	-5.5444467	2.3598521	1.7733283
H	-3.8826880	1.9279653	3.5794145
H	-1.6135684	1.0996924	3.0136817
H	-0.4551305	0.5934401	-0.5619811
H	1.6674849	6.4485153	-0.1600121

Table C.6.: Bz Geometry 1

Atom	x	y	z
H	2.3559866	2.9578717	2.2863705
H	-0.6706673	2.8923590	-0.7778436

Table C.6.: Geometry data of ground state minimum 1 (Geometry 1, see Fig. 5.7) of Bz optimized by DFT. The geometry is given in Cartesian coordinates and units of angstroms.

Table C.7.: Bz Geometry 2

Atom	x	y	z
C	2.6368260	0.8173972	-0.3183687
C	1.4385375	0.1462583	0.0293144
C	0.9164342	-0.8256845	-0.8564337
C	1.5774267	-1.1081712	-2.0624833
C	2.7641988	-0.4346466	-2.3973092
C	3.2944960	0.5302084	-1.5194179
C	0.8076733	0.5435269	1.3322765
O	1.3215065	1.4021006	2.0448688
C	-0.5557059	-0.0266463	1.8194927
O	-0.7187646	-1.3943084	1.4691468
C	-1.6863192	0.8770126	1.3112018
C	-1.8265077	2.1842691	1.8253787
C	-2.8572874	3.0201402	1.3704232
C	-3.7646172	2.5612354	0.3978617
C	-3.6293134	1.2625705	-0.1189139
C	-2.5928089	0.4260622	0.3319536
H	1.1631690	-1.8679691	-2.7461786

Table C.7.: Bz Geometry 2

Atom	x	y	z
H	4.2261083	1.0609382	-1.7783409
H	-1.5774210	-1.6905161	1.8442573
H	-2.4855336	-0.5907417	-0.0797435
H	-4.3338218	0.8952102	-0.8841428
H	-4.5767232	3.2177245	0.0433795
H	-2.9558935	4.0387487	1.7815877
H	-1.1109633	2.5453023	2.5828066
H	-0.4952936	0.0871200	2.9274722
H	3.2808985	-0.6625986	-3.3450626
H	0.0031807	-1.3706228	-0.5823217
H	3.0271072	1.5717720	0.3832450

Table C.7.: Geometry data of ground state minimum 2 (Geometry 2, see Fig. 5.7) of Bz optimized by DFT. The geometry is given in Cartesian coordinates and units of angstroms.

C.4. 2,4,6-Trimethylbenzoin (TMB)

Table C.8.: TMB Geometry 1

Atom	x	y	z
C	-1.594137	0.944079	3.459516
H	-1.256041	0.336485	2.587634
H	-0.861559	0.774617	4.278203
H	-2.574952	0.533633	3.782789
C	-1.694512	2.408171	3.094422
C	-0.576219	3.256870	3.186584
H	0.375174	2.854571	3.577461
C	-2.907280	2.962200	2.639809
H	-3.802945	2.319581	2.576155
C	-3.023080	4.314818	2.274032
H	-4.743090	5.618630	2.543537
H	-4.276157	5.378667	0.835881
H	-5.109066	4.056838	1.724910
C	-0.631168	4.613066	2.809988
C	0.612508	5.469597	2.964391
H	1.228769	5.483707	2.036005
H	1.262344	5.069135	3.771683
H	0.376090	6.524038	3.222078
C	-1.863550	5.136999	2.337503
C	-1.979447	6.583568	1.955476
O	-2.703506	7.370100	2.563093
C	-1.223277	7.138753	0.713703
O	-1.351141	8.533785	0.707633
H	-1.994042	8.721399	1.447467
H	-1.602748	6.706411	-0.104271
C	0.172933	6.760711	0.814613

Table C.8.: TMB Geometry 1

Atom	x	y	z
C	0.612571	5.551959	0.261769
H	-0.094054	4.895646	-0.244052
C	1.960636	5.186952	0.359199
H	2.302612	4.246716	-0.070835
C	2.869061	6.030697	1.009474
H	3.917663	5.746774	1.085260
C	-4.356793	4.867358	1.821124
C	2.413721	7.282619	1.582062
C	0.989796	7.430027	1.351390
H	3.255154	7.798735	2.041994
H	0.647820	8.370263	1.781424

Table C.8.: Geometry data of ground state minimum 1 (Geometry 1, see Fig. 5.7) of TMB optimized by DFT. The geometry is given in Cartesian coordinates and units of angstroms.

Table C.9.: TMB Geometry 2

Atom	x	y	z
C	2.902874	0.289756	-0.077877
H	1.828838	0.335235	-0.361039
H	3.429794	-0.278448	-0.880231
H	2.988303	-0.310316	0.854502
C	3.495539	1.671230	0.091315
C	4.611339	1.885114	0.921098
H	5.033636	1.035723	1.487257
C	2.958772	2.781088	-0.589953
H	2.074701	2.640963	-1.236969
C	3.501779	4.072211	-0.460451
C	2.857179	5.237403	-1.183322
H	2.415976	5.960919	-0.463857
H	2.051360	4.886030	-1.861798
H	3.590617	5.806574	-1.798347
C	5.210149	3.152143	1.060816
C	6.417369	3.308438	1.960365
H	6.458887	4.306078	2.448695
H	7.356017	3.192062	1.374589
H	6.418408	2.539605	2.762891
C	4.649736	4.248855	0.359427
C	5.186770	5.650454	0.531519
O	4.523187	6.528517	1.062324
O	7.169208	4.877341	-0.651296
H	7.977573	5.165789	-1.128706
C	6.556061	6.034321	-0.111530
H	6.409689	6.851311	-0.816529
C	7.477797	6.451506	0.927152
C	8.778538	6.845869	0.591667

Table C.9.: TMB Geometry 2

Atom	x	y	z
C	9.668490	7.248668	1.594532
C	9.257701	7.257104	2.932882
C	7.956960	6.862740	3.268367
C	7.067008	6.459941	2.265502
H	9.098073	6.839307	-0.449379
H	10.680281	7.555426	1.333573
H	9.949956	7.570424	3.712968
H	7.637424	6.869302	4.309412
H	6.055217	6.153183	2.526461

Table C.9.: Geometry data of ground state minimum 2 (Geometry 2, see Fig. 5.7) of TMB optimized by DFT. The geometry is given in Cartesian coordinates and units of angstroms.

C.5. Mesitil (Me)

Table C.10.: Me Geometry 1

Atom	x	y	z
C	1.9375763	-0.2455234	-0.0756153
C	1.1615670	0.0944634	-1.2305779
C	1.7880527	0.0984996	-2.5223646
C	3.1482222	-0.2457709	-2.6188607
C	3.9270112	-0.5771630	-1.4956194
C	3.2992457	-0.5592121	-0.2373117
C	-0.2712125	0.4970875	-1.1277625
O	-0.8922288	1.0724937	-2.0258247
C	1.0619264	0.4405330	-3.8050844
C	5.3798390	-0.9664669	-1.6404060
C	1.3870168	-0.3047799	1.3358502
C	-1.1917616	-0.0278846	-0.0024461
O	-1.2336707	-1.2599510	0.0529576
C	-2.1313326	0.8726989	0.7278848
C	-1.8448281	2.2628490	0.9302708
C	-2.7480783	3.0437672	1.6712203
C	-3.9407419	2.5196752	2.2052528
C	-4.2080136	1.1569845	1.9912986
C	-3.3328584	0.3157259	1.2781844
H	3.6196466	-0.2498672	-3.6173134
H	1.7169008	0.2331645	-4.6791132
H	0.1200328	-0.1365627	-3.9204837
H	0.7577178	1.5086301	-3.8342491
H	5.9600559	-0.7410069	-0.7196798
H	5.4801203	-2.0611387	-1.8303807
H	5.8620671	-0.4428797	-2.4942539

Table C.10.: Me Geometry 1

Atom	x	y	z
H	3.8935866	-0.8033900	0.6605529
H	0.8083611	-1.2400466	1.4963696
H	2.2179259	-0.2837723	2.0729726
H	0.7036943	0.5350265	1.5827720
H	-5.1377238	0.7234856	2.3989022
H	-2.5117851	4.1100416	1.8338966
C	-0.6206772	2.9697390	0.3811666
H	-0.4552922	3.9271039	0.9204603
H	0.3132148	2.3751954	0.4687509
H	-0.7478023	3.2014390	-0.6984240
C	-4.8942304	3.4017262	2.9778193
H	-5.7632096	2.8284016	3.3650260
H	-4.3878503	3.8831397	3.8449627
H	-5.2880675	4.2250634	2.3387671
C	-3.7320768	-1.1354323	1.1227887
H	-4.7657150	-1.2881416	1.5028369
H	-3.6885256	-1.4718245	0.0652311
H	-3.0467305	-1.8124494	1.6759390

Table C.10.: Geometry data of ground state minimum 1 (Geometry 1, see Fig. 5.7) of Me optimized by DFT. The geometry is given in Cartesian coordinates and units of angstroms.

Table C.11.: Me Geometry 2

Atom	x	y	z
C	6.1767461	-1.0571073	-0.3649778
H	6.2584657	-2.1419329	-0.6105236
H	6.7437425	-0.5037112	-1.1447372
H	6.6872442	-0.9035775	0.6100359
C	4.7296667	-0.6219619	-0.3128637
C	4.0504560	-0.2207973	-1.4796769
H	4.5974209	-0.1941356	-2.4384278
C	4.0135923	-0.6137983	0.8978154
H	4.5346617	-0.8827904	1.8334616
C	2.6982259	0.1646836	-1.4658048
C	2.0484437	0.6262472	-2.7547042
H	1.1142319	0.0660264	-2.9789008
H	1.7685435	1.7001288	-2.7000404
H	2.7384834	0.4882954	-3.6134817
C	2.6516245	-0.2631095	0.9637925
C	1.9708599	-0.2522525	2.3178084
H	1.2170362	0.5604416	2.4039543
H	1.4459871	-1.2128866	2.5102154
H	2.7152254	-0.0999189	3.1285532
C	1.9827666	0.1097809	-0.2335721
C	0.5350774	0.5272105	-0.2301900
O	0.1599773	1.6232788	-0.6288901
C	-0.5335232	-0.5179151	0.2514655
O	-0.1644522	-1.6165472	0.6476757
C	-1.9820556	-0.0997472	0.2443785
C	-2.7000860	-0.1213442	1.4763321
C	-2.0444372	-0.5328100	2.7791026
H	-1.1311956	0.0645280	2.9952136

Table C.11.: Me Geometry 2

Atom	x	y	z
H	-1.7291761	-1.5978448	2.7507907
H	-2.7434394	-0.3981595	3.6311216
C	-2.6476611	0.2357023	-0.9639820
C	-1.9576690	0.1994727	-2.3126613
H	-1.2016972	-0.6134063	-2.3773728
H	-1.4334854	1.1576285	-2.5197363
H	-2.6953779	0.0304570	-3.1260759
C	-4.0134271	0.5783036	-0.9127176
H	-4.5323137	0.8206269	-1.8567379
C	-4.0546515	0.2527777	1.4750492
H	-4.6034638	0.2523200	2.4333272
C	-4.7341972	0.6120923	0.2935230
C	-6.1911793	1.0151119	0.3324604
H	-6.3242220	1.9899423	0.8557562
H	-6.6145857	1.1222997	-0.6886922
H	-6.8061830	0.2683979	0.8829339

Table C.11.: Geometry data of ground state minimum 2 (Geometry 2, see Fig. 5.7) of Me optimized by DFT. The geometry is given in Cartesian coordinates and units of angstroms.

C.6. 7-Diethylamino-3-thenoylcoumarin (DETC)

Table C.12.: Minimum geometry DETC (S_0)

Atom	x	y	z
C	-0.6033587	0.5969620	6.3865971
S	-0.0741050	-0.8208995	5.5708477
C	-0.0710637	0.0143124	4.0273453
C	-0.4770663	1.3299532	4.1828578
C	-0.7819191	1.6609276	5.5295086
C	0.2804355	-0.8342933	2.8713329
O	0.4895407	-2.0277433	3.0712907
C	0.3084405	-0.3494331	1.4425316
C	0.1275956	-1.3202658	0.4841972
C	0.1451947	-1.0390902	-0.9077437
C	0.3795123	0.2978055	-1.3013158
O	0.6002677	1.2476309	-0.3642102
C	0.6106149	1.0215978	1.0176796
C	0.4061143	0.6864090	-2.6362258
C	0.1961787	-0.2657765	-3.6644549
C	-0.0361974	-1.6239461	-3.2691282
C	-0.0585802	-1.9864176	-1.9373684
N	0.2119234	0.0983728	-4.9901758
C	0.3411061	1.4953791	-5.4011842
C	1.7880528	1.9758125	-5.5511274
O	0.8846278	1.9744940	1.7016968
C	0.1016339	-0.8841608	-6.0666576
C	-1.3384829	-1.2106141	-6.4740652
H	-0.0344229	-2.3446908	0.8299352
H	-0.2152877	-2.3911305	-4.0200362
H	-0.2446558	-3.0288801	-1.6647534

Table C.12.: Minimum geometry DETC (S_0)

Atom	x	y	z
H	0.6119525	1.7341040	-2.8438690
H	0.6489708	-0.4834916	-6.9353298
H	0.6390673	-1.8023688	-5.7823155
H	-1.9054480	-1.6470938	-5.6370285
H	-1.8728929	-0.3053737	-6.8047455
H	-1.3487045	-1.9316585	-7.3077570
H	-0.1871051	1.6067751	-6.3622098
H	-0.2046654	2.1355022	-4.6909133
H	1.8103765	3.0266193	-5.8828439
H	2.3368602	1.9066268	-4.5994781
H	2.3319224	1.3745632	-6.2975695
H	-0.5228605	2.0363141	3.3582090
H	-1.1132812	2.6495584	5.8522624
H	-0.7574302	0.5817326	7.4658488

Table C.12.: Geometry data of the DETC ground state optimized by DFT (see Fig. 6.2). The geometry is given in Cartesian coordinates and units of angstroms.

Table C.13.: Minimum geometry of DETC (T_1)

Atom	x	y	z
C	-0.4593235	0.6441253	6.4285713
S	0.0246619	-0.7718707	5.5770452
C	-0.0847512	0.0583194	4.0345488
C	-0.4959550	1.3694271	4.2174086
C	-0.7097358	1.7019536	5.5826206
C	0.1818314	-0.8022816	2.8645867
O	0.3393121	-2.0176148	3.0647002
C	0.2199086	-0.3573674	1.4547352
C	0.0743772	-1.3981116	0.4579271
C	0.1080052	-1.0914408	-0.8962576
C	0.3382053	0.2722351	-1.3021879
O	0.5509885	1.2340120	-0.3602541
C	0.5356697	1.0003842	1.0140383
C	0.3756913	0.6585332	-2.6200134
C	0.1884780	-0.2939346	-3.6754088
C	-0.0358695	-1.6554827	-3.2920677
C	-0.0738787	-2.0348098	-1.9682740
N	0.2219319	0.0943002	-4.9986774
C	0.3339020	1.4989592	-5.3935744
C	1.7800016	1.9932781	-5.5252516
O	0.8042547	1.9552257	1.7035617
C	0.1469431	-0.8732629	-6.0929325
C	-1.2853811	-1.2053756	-6.5295188
H	-0.0778980	-2.4180716	0.8090438
H	-0.1981740	-2.4213403	-4.0509210
H	-0.2547105	-3.0815590	-1.7086863
H	0.5739107	1.7109777	-2.8178345
H	0.7052764	-0.4487620	-6.9447082

Table C.13.: Minimum geometry of DETC (T_1)

Atom	x	y	z
H	0.6876695	-1.7894892	-5.8065388
H	-1.8607499	-1.6621100	-5.7058668
H	-1.8229187	-0.2974061	-6.8562853
H	-1.2697591	-1.9147213	-7.3763960
H	-0.1859655	1.6095251	-6.3606870
H	-0.2248478	2.1231884	-4.6791436
H	1.7907962	3.0483517	-5.8525654
H	2.3172101	1.9244708	-4.5640141
H	2.3383275	1.3995592	-6.2707970
H	-0.6257362	2.0735928	3.3989047
H	-1.0322021	2.6891545	5.9235999
H	-0.5366355	0.6295627	7.5172044

Table C.13.: Geometry data of the DETC T_1 state optimized by DFT. The geometry is given in Cartesian coordinates and units of angstroms.

C.7. Isopropylthioxanthone (ITX)

Table C.14.: Minimum geometry of 2-isopropyl-9-thioxanthenone (S_0)

Atom	x	y	z
C	-1.1301250	1.7687638	0.0000000
C	-1.1262382	0.3768478	0.0000000
C	0.0632124	-0.3828546	0.0000000
C	1.2966923	0.2974121	0.0000000
C	1.3109229	1.7067122	0.0000000
C	0.1218984	2.4225378	0.0000000
C	-0.0758204	-1.8667479	0.0000000
C	2.4688988	-2.2065632	0.0000000
C	1.1519552	-2.7134151	0.0000000
C	0.9701038	-4.1131524	0.0000000
H	-0.0593795	-4.4764798	0.0000000
C	2.0499572	-4.9843976	0.0000000
C	3.3562966	-4.4668212	0.0000000
C	3.5654032	-3.0933273	0.0000000
H	-2.0622538	-0.1865796	0.0000000
H	2.2676530	2.2363160	0.0000000
H	0.1687559	3.5150396	0.0000000
H	1.8862733	-6.0648361	0.0000000
H	4.2161232	-5.1420826	0.0000000
H	4.5824248	-2.6918583	0.0000000
O	-1.1850701	-2.3861873	0.0000000
S	2.8678601	-0.4936072	0.0000000
C	-2.4363363	2.5523413	0.0000000
H	-3.2499089	1.8061929	0.0000000
C	-2.5948182	3.4058216	1.2711435
H	-3.5742743	3.9117903	1.2802840

Table C.14.: Minimum geometry of 2-isopropyl-9-thioxanthenone (S_0)

Atom	x	y	z
H	-2.5218092	2.7881842	2.1802160
H	-1.8187479	4.1875772	1.3340094
C	-2.5948182	3.4058216	-1.2711435
H	-2.5218092	2.7881842	-2.1802160
H	-3.5742743	3.9117903	-1.2802840
H	-1.8187479	4.1875772	-1.3340094

Table C.14.: Geometry data of the 2-isopropyl-9-thioxanthenone ground state optimized by DFT (see Fig. 6.2). The geometry is given in Cartesian coordinates and units of angstroms.

Table C.15.: Minimum geometry of 4-isopropyl-9-thioxanthenone (S_0)

Atom	x	y	z
S	0.4513139	1.0696338	0.0000000
C	-0.6052766	-0.3528714	0.0000000
C	-0.0807712	-1.6778642	0.0000000
C	1.3731032	-2.0387654	0.0000000
C	2.3978852	-0.9589924	0.0000000
C	2.0855368	0.4221271	0.0000000
C	-2.0202790	-0.1257429	0.0000000
C	-2.8543353	-1.2585175	0.0000000
C	-2.3527745	-2.5734779	0.0000000
C	-0.9762248	-2.7743096	0.0000000
C	3.7617307	-1.3458382	0.0000000
C	4.7849434	-0.3980882	0.0000000
C	4.4598688	0.9769569	0.0000000

Table C.15.: Minimum geometry of 4-isopropyl-9-thioxanthenone (S_0)

Atom	x	y	z
C	3.1242412	1.3865146	0.0000000
H	-3.0463978	-3.4300562	0.0000000
O	1.7213632	-3.2256601	0.0000000
H	3.9677401	-2.4277966	0.0000000
H	-0.5260738	-3.7790997	0.0000000
C	-2.6999514	1.2511259	0.0000000
H	-3.9465670	-1.0987851	0.0000000
H	5.8397327	-0.7176867	0.0000000
H	5.2593465	1.7366185	0.0000000
H	2.8712268	2.4600245	0.0000000
H	-3.7897953	1.0249684	0.0000000
C	-2.4425403	2.0700225	1.2859069
C	-2.4425403	2.0700225	-1.2859069
H	-3.0732231	2.9870890	1.2823949
H	-2.7013075	1.4794518	2.1920646
H	-1.3834037	2.3914654	1.3833908
H	-3.0732231	2.9870890	-1.2823949
H	-1.3834037	2.3914654	-1.3833908
H	-2.7013075	1.4794518	-2.1920646

Table C.15.: Geometry data of the 4-isopropyl-9-thioxanthenone ground state optimized by DFT. The geometry is given in Cartesian coordinates and units of angstroms.

Table C.16.: Minimum geometry of 2-isopropyl-9-thioxanthenone T_1

Atom	x	y	z
C	-1.1272596	1.7744342	0.0016374
C	-1.1255388	0.3804924	0.0014181
C	0.0742778	-0.3822884	-0.0005267
C	1.3121319	0.3154811	-0.0007857
C	1.3125056	1.7145035	0.0002786
C	0.1167553	2.4369775	0.0008947
C	0.0200257	-1.8196228	-0.0030699
C	2.4881557	-2.2130187	-0.0009478
C	1.1526165	-2.7064581	-0.0024328
C	0.9613468	-4.1154623	-0.0024858
H	-0.0599978	-4.5065454	-0.0038548
C	2.0414692	-4.9903970	-0.0003277
C	3.3512897	-4.4903913	0.0020842
C	3.5644757	-3.1077037	0.0016469
H	-2.0790000	-0.1568518	0.0021505
H	2.2685751	2.2480750	0.0004800
H	0.1614673	3.5299912	0.0008153
H	1.8637004	-6.0701898	-0.0002431
H	4.2070971	-5.1715204	0.0043929
H	4.5855255	-2.7127073	0.0032256
O	-1.1856584	-2.3795593	-0.0061619
S	2.8879423	-0.4891802	-0.0031701
C	-2.4402101	2.5493737	0.0015366
H	-3.2544356	1.8014813	0.0031887
C	-2.6029215	3.4053000	1.2722016
H	-3.5881765	3.9067917	1.2796834
H	-2.5264493	2.7866549	2.1838579
H	-1.8276620	4.1914668	1.3333244

Table C.16.: Minimum geometry of 2-isopropyl-9-thioxanthenone T_1

Atom	x	y	z
C	-2.6043709	3.4012656	-1.2716374
H	-2.5286666	2.7797122	-2.1813807
H	-3.5897242	3.9025547	-1.2797617
H	-1.8292861	4.1873407	-1.3360307

Table C.16.: Geometry data of the T_1 state of 2-isopropyl-9-thioxanthenone optimized by DFT. The geometry is given in Cartesian coordinates and units of angstroms.

Bibliography

[1] G. Moad, *The chemistry of radical polymerization*, 2. fully rev. ed., Elsevier, Amsterdam **2006**.

[2] J. Warnatz, U. Maas, R. W. Dibble, *Combustion : Physical and Chemical Fundamentals, Modelling and Simulation, Experiments, Pollutant Formation*, Springer, Berlin **1996**.

[3] J. I. Steinfeld, J. S. Francisco, W. L. Hase, *Chemical kinetics and dynamics*, 2 ed., Prentice Hall, Upper Saddle River **1999**.

[4] K. H. Cheeseman, T. F. Slater, *Br. Med. Bull.* **1993**, *49*, 481.

[5] M. D. Lechner, K. Gehrke, E. H. Nordmeier, *Makromolekulare Chemie*, 3 ed., Birkhäuser **2003**.

[6] W. M. Haynes (Ed.), *2011 - 2012*, CRC Handbook of chemistry and physics, 92 ed., CRC Press, Boca Raton, Fla. **2011**.

[7] D. Voll, T. Junkers, C. Barner-Kowollik, *Macromolecules* **2011**, *44*, 2542.

[8] J. Fischer, M. Wegener, *Laser & Photon. Rev.* **2012**, 1.

[9] M. Kasha, *Discuss. Faraday Soc.* **1950**, *9*, 14.

[10] M. Klessinger, J. Michl, *Excited states and photochemistry of organic molecules*, 1 ed., Wiley, New York **1995**.

[11] R. Cox, *J. Photochem.* **1974**, *3*, 175.

[12] P. J. Graedel, Thomas E. ; Crutzen, *Chemie der Atmosphäre: Bedeutung für Klima und Umwelt*, Spektrum, Heidelberg **1994**.

[13] S. W. Benson, *Thermochemical kinetics: methods for the estimation of thermochemical data and rate parameters*, Wiley, New York **1968**.

[14] B. P. Fingerhut, C. F. Sailer, J. Ammer, E. Riedle, R. de Vivie-Riedle, *J. Phys. Chem. A* **2012**, DOI: 10.1021/jp300986t.

[15] M. Gomberg, *J. Am. Chem. Soc.* **1900**, *22*, 757.

[16] S. A. Rice, *Diffusion-limited reactions*, Comprehensive chemical kinetics, Elsevier **1985**.

[17] M. von Smoluchowski, *Z. Phys. Chem.* **1917**, *92*, 129.

[18] J. Fischer, M. Wegener, *Adv. Mater.* **2012**, *24*, OP65.

[19] T. H. Maiman, *Nature* **1960**, *187*, 493.

[20] D. A. Skoog, F. J. Holler, S. R. Crouch, *Principles of instrumental analysis*, 6 ed., Thomson, Brooks/Cole, Belmont **2007**.

[21] F. J. McClung, R. W. Hellwarth, *J. Appl. Phys.* **1962**, *33*, 828.

[22] E. P. Ippen, C. V. Shank, *Appl. Phys. Lett.* **1975**, *27*, 488.

[23] M. H. Chergui (Ed.), *Ultrafast phenomena XVII : proceedings of the 17th International Conference at The Silvertree Hotel and Snowmass Conference Center, Snowmass, Colorado, United States, July 18 - 23, 2010*, vol. 17, Oxford Univ. Press, Oxford **2011**.

[24] A. H. Zewail., *Angew. Chem., Int. Ed.* **2000**, *39*, 2586.

[25] W. Demtröder, *Laserspektroskopie: Grundlagen und Techniken*, 5 ed., Springer-Verlag Berlin Heidelberg, Berlin, Heidelberg **2007**.

[26] A. H. Zewail, *J. Phys. Chem.* **1993**, *97*, 12427.

[27] R. W. Boyd, *Nonlinear optics*, 3 ed., Academic Press, Burlington **2008**.

[28] J.-C. Diels, W. Rudolph, *Ultrashort Laser Pulse Phenomena*, 2 ed., Academic Press **2006**.

[29] C. H. Rullière (Ed.), *Femtosecond laser pulses: principles and experiments*, 2 ed., Springer, New York **2005**.

[30] D. Meschede, *Optik, Licht und Laser*, 3 ed., Teubner, Wiesbaden **2008**.

[31] M. D. Levenson, S. S. Kano, *Introduction to nonlinear laser spectroscopy*, rev. ed., Acad. Press, Boston **1988**.

[32] R. R. Alfano, S. L. Shapiro, *Phys. Rev. Lett.* **1970**, *24*, 592.

[33] A. Zheltikov, *Appl. Phys. B: Lasers Opt.* **2003**, *77*, 143.

[34] R. R. Alfano (Ed.), *The supercontinuum laser source: fundamentals with updated references*, 2 ed., Springer, New York, NY **2006**.

[35] P. G. Kryukov, *Quantum Electron.* **2001**, *31*, 95.

[36] D. Strickland, G. Mourou, *Opt. Commun.* **1985**, *55*, 447 .

[37] P. Maine, D. Strickland, P. Bado, M. Pessot, G. Mourou, *IEEE J. Quantum Electron.* **1988**, *24*, 398.

[38] M. Klinger, *Ultrakurzzeitdynamik ausgewählter Germaniumcluster, mehrkerniger Seltenerdverbindungen sowie quasi monodisperser (9,7)-Kohlenstoffnanoröhren in Lösung*, Ph.D. thesis, Karlsruhe Institute of Technology (KIT), Karlsruhe **2011**.

[39] H. A. Haus, J. G. Fujimoto, E. P. Ippen, *IEEE J. Quantum Electron.* **1992**, *28*, 2086.

[40] K. Tamura, C. R. Doerr, L. E. Nelson, H. A. Haus, E. P. Ippen, *Opt. Lett.* **1994**, *19*, 46.

[41] G. Lenz, K. Tamura, H. A. Haus, E. P. Ippen, *Opt. Lett.* **1995**, *20*, 1289.

[42] Clark MXR, *CPA-2001 User Manual*, 4.1 ed. **2002**.

[43] O. Schalk, *Modell zur Beschreibung optischer Anisotropie sowie Ultrakurzzeitspektroskopie an Porphyrinen und perchloriertem Cycloheptatrien*, Ph.D. thesis, Universität Karlsruhe (TH) **2008**.

[44] I. N. Duling, T. Norris, T. Sizer, P. Bado, G. A. Mourou, *J. Opt. Soc. Am. B* **1985**, *2*, 616.

[45] P. Bado, M. Bouvier, J. S. Coe, *Opt. Lett.* **1987**, *12*, 319.

[46] D. Du, J. Squier, S. Kane, G. Korn, G. Mourou, C. Bogusch, C. T. Cotton, *Opt. Lett.* **1995**, *20*, 2114.

[47] T. Wilhelm, J. Piel, E. Riedle, *Opt. Lett.* **1997**, *22*, 1494.

[48] E. Riedle, M. Beutter, S. Lochbrunner, J. Piel, S. Schenkl, S. Spörlein, W. Zinth, *Appl. Phys. B: Lasers Opt.* **2000**, *71*, 457.

[49] J. Piel, M. Beutter, E. Riedle, *Opt. Lett.* **2000**, *25*, 180.

[50] H. Brands, *Ultrakurzzeitdynamik von Fulleriden in Lösung und suspendierten, längenselektierten Kohlenstoffnanoröhren*, Ph.D. thesis, Universität Karlsruhe (TH) **2008**.

[51] A. Stolow, J. G. Underwood, in *Advances in Chemical Physics*, vol. 139, Wiley **2008**, 497.

[52] W. Kutzelnigg, *Einführung in die Theoretische Chemie*, Wiley, Weinheim **2002**.

[53] F. Jensen, *Introduction to computational chemistry*, 2 ed., Wiley, Chichester **2007**.

[54] T. Helgaker, P. Jørgensen, J. Olsen, *Molecular electronic-structure theory*, Wiley, Chichester **2000**.

[55] M. Born, J. R. Oppenheimer, *Anal. Phys.* **1927**, *389*, 457.

[56] G. A. Worth, L. S. Cederbaum, *Annu. Rev. Phys. Chem.* **2004**, *55*, 127.

[57] W. H. Domcke (Ed.), *Conical intersections: electronic structure, dynamics & spectroscopy*, Advanced series in physical chemistry ; 15, World Scientific, Singapore **2004**.

[58] O. Christiansen, H. Koch, P. Jørgensen, *Chem. Phys. Lett.* **1995**, *243*, 409.

[59] O. Christiansen, H. Koch, P. Jørgensen, *J. Chem. Phys.* **1995**, *103*, 7429.

[60] O. Christiansen, P. Jørgensen, C. Hättig, *Int. J. Quantum Chem.* **1998**, *68*, 1.

[61] C. Hättig, in *Computational nanoscience: do it yourself!*, *NIC series*, vol. 31, J. Grotendorst, ed., NIC, Jülich, 245, **2006**.

[62] J. Olsen, P. Jørgensen, in *Modern Electronic Structure Theory*, vol. 2, D. R. Yarkony, ed., chap. 13, World Scientific **1995**, 857.

[63] P. Hohenberg, W. Kohn, *Phys. Rev.* **1964**, *136*, B864.

[64] W. Kohn, L. J. Sham, *Phys. Rev.* **1965**, *140*, A1133.

[65] S. H. Vosko, L. Wilk, M. Nusair, *Can. J. Phys.* **1980**, *58*, 1200.

[66] A. D. Becke, *Phys. Rev. A* **1988**, *38*, 3098.

[67] C. Lee, W. Yang, R. G. Parr, *Phys. Rev. B* **1988**, *37*, 785.

[68] J. P. Perdew, *Phys. Rev. B* **1986**, *33*, 8822.

[69] A. D. Becke, *J. Chem. Phys.* **1993**, *98*, 5648.

[70] P. J. Stephens, F. J. Devlin, C. F. Chabalowski, M. J. Frisch, *J. Phys. Chem.* **1994**, *98*, 11623.

[71] F. Furche, *J. Chem. Phys.* **2001**, *114*, 5982.

[72] F. Furche, *Dichtefunktionalmethoden für elektronisch angeregte Moleküle: Theorie - Implementierung - Anwendung*, Ph.D. thesis, Universität Karlsruhe (TH), Göttingen **2002**.

[73] B. Engels, M. Hanrath, C. Lennartz, *Computers & Chemistry* **2001**, *25*, 15.

[74] B. G. Levine, C. Ko, J. Quenneville, T. J. Martínez, *Mol. Phys.* **2006**, *104*, 1039.

[75] M. W. Schmidt, M. S. Gordon, *Annu. Rev. Phys. Chem.* **1998**, *49*, 233.

[76] P. G. Szalay, T. Müller, G. Gidofalvi, H. Lischka, R. Shepard, *Chem. Rev.* **2012**, *112*, 108.

[77] V. V. Ivanov, D. I. Lyakh, L. Adamowicz, *Annu. Rep. Prog. Chem., Sect. C: Phys. Chem.* **2011**, *107*, 169.

[78] P. Pulay, *Int. J. Quantum Chem.* **2011**, *111*, 3273.

[79] A. K, B. O. Roos, in *Modern Electronic Structure Theory*, vol. 1, D. R. Yarkony, ed., World Scientific **1995**, 55.

[80] I. Z. Kozma, P. Baum, S. Lochbrunner, E. Riedle, *Opt. Express* **2003**, *11*, 3110.

[81] S. Pedersen, A. H. Zewail, *Mol. Phys.* **1996**, *89*, 1455.

[82] I. Z. Kozma, P. Krok, E. Riedle, *J. Opt. Soc. Am. B* **2005**, *22*, 1479.

[83] F. Strauss, L. Kollek, W. Heyn, *Ber. deut. chem. Ges. B* **1930**, *63*, 1868.

[84] S. Lochbrunner, J. J. Larsen, J. P. Shaffer, M. Schmitt, T. Schultz, J. G. Underwood, A. Stolow, *J. Electron Spectrosc. Relat. Phenom.* **2000**, *112*, 183.

[85] G. K. Jarvis, M. Evans, C. Y. Ng, K. Mitsuke, *J. Chem. Phys.* **1999**, *111*, 3058.

[86] A. R. Gray, R. C. Fuson, *J. Am. Chem. Soc.* **1934**, *56*, 739.

[87] H. H. Weinstock, R. C. Fuson, *J. Am. Chem. Soc.* **1936**, *58*, 1986.

[88] F. Günzler, E. H. H. Wong, S. P. S. Koo, T. Junkers, C. Barner-Kowollik, *Macromolecules* **2009**, *42*, 1488.

[89] N. Nudelman, H. Schulz, *J. Chem. Res. (S)* **1999**, 422.

[90] C. J. Woltermann, T. L. Rathman, *PharmaChem* **2003**, *2*, 4.

[91] J. Fischer, *Three-dimensional optical lithography beyond the difraction limit*, Ph.D. thesis, Karlsruhe Institute of Technology (KIT) **2012**.

[92] J. Fischer, G. von Freymann, M. Wegener, *Adv. Mater.* **2010**, *22*, 3578.

[93] R. Ahlrichs, M. Bär, M. Häser, H. Horn, C. Kölmel, *Chem. Phys. Lett.* **1989**, *162*, 165.

[94] O. Treutler, R. Ahlrichs, *J. Chem. Phys.* **1995**, *102*, 346.

[95] K. Eichkorn, O. Treutler, H. Öhm, M. Häser, R. Ahlrichs, *Chem. Phys. Lett.* **1995**, *240*, 283.

[96] K. Eichkorn, O. Treutler, H. Öhm, M. Häser, R. Ahlrichs, *Chem. Phys. Lett.* **1995**, *242*, 652.

[97] K. Eichkorn, F. Weigend, O. Treutler, R. Ahlrichs, *Theor. Chim. Acta* **1997**, *97*, 119.

[98] A. Schäfer, H. Horn, R. Ahlrichs, *J. Chem. Phys.* **1992**, *97*, 2571.

[99] F. Weigend, R. Ahlrichs, *Phys. Chem. Chem. Phys.* **2005**, *7*, 3297.

[100] F. Weigend, *Phys. Chem. Chem. Phys.* **2006**, *8*, 1057.

[101] *USER'S MANUAL for Turbomole Version 6.3.*

[102] C. Hättig, F. Weigend, *J. Chem. Phys.* **2000**, *113*, 5154.

[103] C. Hättig, A. Köhn, *J. Chem. Phys.* **2002**, *117*, 6939.

[104] A. Köhn, C. Hättig, *J. Chem. Phys.* **2003**, *119*, 5021.

[105] R. A. Kendall, T. H. Dunning, R. J. Harrison, *J. Chem. Phys.* **1992**, *96*, 6796.

[106] F. Weigend, A. Kohn, C. Hättig, *J. Chem. Phys.* **2002**, *116*, 3175.

[107] F. Furche, D. Rappoport, in *Computational and Theoretical Chemistry*, vol. 16, M. Olivucci, ed., Elsevier, Amsterdam **2005**.

[108] R. Bauernschmitt, R. Ahlrichs, *Chem. Phys. Lett.* **1996**, *256*, 454.

[109] R. Bauernschmitt, R. Ahlrichs, *J. Chem. Phys.* **1996**, *104*, 9047.

[110] H. Weiss, R. Ahlrichs, M. Haser, *J. Chem. Phys.* **1993**, *99*, 1262.

[111] D. Jacquemin, V. Wathelet, E. A. Perpète, C. Adamo, *J. Chem. Theory Comput.* **2009**, *5*, 2420.

[112] F. Trani, G. Scalmani, G. Zheng, I. Carnimeo, M. J. Frisch, V. Barone, *J. Chem. Theory Comput.* **2011**, *7*, 3304.

[113] F. Weigend, M. Häser, H. Patzelt, R. Ahlrichs, *Chem. Phys. Lett.* **1998**, *294*, 143.

[114] A. Hellweg, C. Hättig, S. Höfener, W. Klopper, *Theor. Chim. Acta* **2007**, *117*, 587.

[115] H. Lischka, R. Shepard, F. B. Brown, I. Shavitt, *Int. J. Quantum Chem.* **1981**, *20*, 91.

[116] R. Shepard, I. Shavitt, R. M. Pitzer, D. C. Comeau, M. Pepper, H. Lischka, P. G. Szalay, R. Ahlrichs, F. B. Brown, J.-G. Zhao, *Int. J. Quantum Chem.* **1988**, *34*, 149.

[117] H. Lischka, R. Shepard, R. M. Pitzer, I. Shavitt, M. Dallos, T. Muller, P. G. Szalay, M. Seth, G. S. Kedziora, S. Yabushita, Z. Zhang, *Phys. Chem. Chem. Phys.* **2001**, *3*, 664.

[118] H. Lischka, T. Müller, P. G. Szalay, I. Shavitt, R. M. Pitzer, R. Shepard, *WIREs Comput. Mol. Sci.* **2011**, *1*, 191.

[119] W. J. Hehre, R. Ditchfield, J. A. Pople, *J. Chem. Phys.* **1972**, *56*, 2257.

[120] M. J. Frisch, J. A. Pople, J. S. Binkley, *J. Chem. Phys.* **1984**, *80*, 3265.

[121] S. Rayne, K. Forest, K. J. Friesen, *Environ. Int.* **2009**, *35*, 425.

[122] D. Raftery, M. Iannone, C. Phillips, R. Hochstrasser, *Chem. Phys. Lett.* **1993**, *201*, 513.

[123] C. G. Elles, M. J. Cox, G. L. Barnes, F. F. Crim, *J. Phys. Chem. A* **2004**, *108*, 10973.

[124] B. Halliwell, *Mutat. Res., Genet. Toxicol. Environ. Mutagen.* **1999**, *443*, 37.

[125] N. Yamamoto, T. Kajikawa, H. Sato, H. Tsubomura, *J. Am. Chem. Soc.* **1969**, *91*, 265.

[126] J. M. Bossy, R. E. Buehler, M. Ebert, *J. Am. Chem. Soc.* **1970**, *92*, 1099.

[127] J. E. Chateauneuf, *Chem. Phys. Lett.* **1989**, *164*, 577.

[128] L. C. T. Shoute, P. Neta, *J. Phys. Chem.* **1990**, *94*, 2447.

[129] L. C. T. Shoute, P. Neta, *J. Phys. Chem.* **1990**, *94*, 7181.

[130] Z. B. Alfassi, R. E. Huie, J. P. Mittal, P. Neta, L. C. T. Shoute, *J. Phys. Chem.* **1993**, *97*, 9120.

[131] J. E. Chateauneuf, *J. Org. Chem.* **1999**, *64*, 1054.

[132] L. Sheps, A. C. Crowther, C. G. Elles, F. F. Crim, *J. Phys. Chem. A* **2005**, *109*, 4296.

[133] S. L. Carrier, T. J. Preston, M. Dutta, A. C. Crowther, F. F. Crim, *J. Phys. Chem. A* **2010**, *114*, 1548.

[134] F. Graf, H. H. Günthard, *Chem. Phys. Lett.* **1970**, *7*, 25.

[135] F. Graf, H. H. Günthard, *Chem. Phys. Lett.* **1971**, *8*, 395.

[136] P. Bachmann, F. Graf, H. Günthard, *Chem. Phys.* **1975**, *9*, 41.

[137] Safety data sheet of cyclohexane, downloaded from Carl Roth 2012.07.18.

[138] Safety data sheet of isopropanol, downloaded from Carl Roth 2012.07.18.

[139] Safety data sheet of chloroform, downloaded from Carl Roth 2012.07.18.

[140] E. S. Snow, F. K. Perkins, E. J. Houser, S. C. Badescu, T. L. Reinecke, *Science* **2005**, *307*, 1942.

[141] Safety data sheet of trichloroethanol, downloaded from Sigma Aldrich 2012.07.18.

[142] S. S. Krishnamurthy, S. Soundararajan, *J. Phys. Chem.* **1969**, *73*, 4036.

[143] N. S. Angerman, R. B. Jordan, *J. Chem. Phys.* **1971**, *54*, 837.

[144] O. Schalk, A.-N. Unterreiner, *J. Phys. Chem. A* **2007**, *111*, 3231.

[145] L. W. Pickett, M. Muntz, E. M. McPherson, *J. Am. Chem. Soc.* **1951**, *73*, 4862.

[146] G. Dey, K. Pushpa, *Res. Chem. Intermed.* **2006**, *32*, 725.

[147] P. C. Simon, D. Gillotay, N. Vanlaethem-Meuree, J. Wisemberg, *J. Atmos. Chem.* **1988**, *7*, 107.

[148] R. B. Woodward, R. Hoffmann, *Angew. Chem., Int. Ed.* **1969**, *8*, 781.

[149] J. I. Brauman, L. E. Ellis, E. E. van Tamelen, *J. Am. Chem. Soc.* **1966**, *88*, 846.

[150] E. E. van Tamelen, J. I. Brauman, L. E. Ellis, *J. Am. Chem. Soc.* **1970**, *93*, 6145.

[151] G. D. Andrews, J. E. Baldwin, *J. Am. Chem. Soc.* **1977**, *99*, 4851.

[152] G. R. Liebling, H. M. McConnell, *J. Chem. Phys.* **1965**, *42*, 3931.

[153] L. W. Pickett, E. Paddock, E. Sackter, *J. Am. Chem. Soc.* **1941**, *63*, 1073.

[154] R. P. Frueholz, W. M. Flicker, O. A. Mosher, A. Kuppermann, *J. Chem. Phys.* **1979**, *70*, 2003.

[155] R. McDiarmid, A. Sabljić, J. P. Doering, *J. Chem. Phys.* **1985**, *83*, 2147.

[156] A. Sabljić, R. McDiarmid, *J. Chem. Phys.* **1990**, *93*, 3850.

[157] W. Yi, A. Chattopadhyay, R. Bersohn, *J. Chem. Phys.* **1991**, *94*, 5994.

[158] N. H. Werstiuk, J. Ma, J. B. Macaulay, A. G. Fallis, *Can. J. Chem.* **1992**, *70*, 2798.

[159] W. Fuß, W. E. Schmid, S. Trushin, *Chem. Phys.* **2005**, *316*, 225.

[160] F. Rudakov, P. M. Weber, *Chem. Phys. Lett.* **2009**, *470*, 187.

[161] O. Schalk, A. E. Boguslavskiy, A. Stolow, *J. Phys. Chem. A* **2010**, *114*, 4058.

[162] F. Rudakov, P. M. Weber, *J. Phys. Chem. A* **2010**, *114*, 4501.

[163] M. Z. Zgierski, F. Zerbetto, *Chem. Phys. Lett.* **1991**, *179*, 131.

[164] M. Z. Zgierski, F. Zerbetto, *J. Chem. Phys.* **1993**, *99*, 3721.

[165] L. Serrano-Andres, M. Merchan, I. Nebot-Gil, B. O. Roos, M. Fulscher, *J. Am. Chem. Soc.* **1993**, *115*, 6184.

[166] T. Kovar, H. Lischka, *J. Mol. Struct.* **1994**, *303*, 71.

[167] H. Nakano, T. Tsuneda, T. Hashimoto, K. Hirao, *J. Chem. Phys.* **1996**, *104*, 2312.

[168] J. Wan, M. Ehara, M. Hada, H. Nakatsuji, *J. Chem. Phys.* **2000**, *113*, 5245.

[169] Y. J. Bomble, K. W. Sattelmeyer, J. F. Stanton, J. Gauss, *J. Chem. Phys* **2004**, *121*, 5236.

[170] S. Tokura, T. Tsuneda, K. Hirao, *J. Theor. Comput. Chem.* **2006**, *5*, 925.

[171] T. S. Kuhlman, W. J. Glover, T. Mori, K. B. Møller, T. Martinez, *Faraday Discuss.* **2012**, *157*, 193.

[172] B. S. Hudson, B. E. Kohler, K. Schulten, in *Excited States*, E. C. Lim, ed., Academic Press **1982**.

[173] M. Aoyagi, Y. Osamura, *J. Am. Chem. Soc.* **1989**, *111*, 470.

[174] M. Olivucci, I. N. Ragazos, F. Bernardi, M. A. Robb, *J. Am. Chem. Soc.* **1993**, *115*, 3710.

[175] P. Celani, B. F, M. Olivucci, M. A. Robb, *J. Chem. Phys.* **1995**, *102*, 5733.

[176] F. Bernardi, M. Olivucci, M. A. Robb, *Chem. Soc. Rev.* **1996**, *25*, 321.

[177] W. Fuß, S. Lochbrunner, A. M. Müller, T. Schirkarski, W. E. Schmid, S. A. Trushin, *Chem. Phys.* **1998**, *232*, 161.

[178] D. R. Yarkony, *J. Phys. Chem. A* **2001**, *105*, 6277.

[179] V. Sundström, *Annu. Rev. Phys. Chem.* **2008**, *59*, 53.

[180] O. Schalk, A. E. Boguslavskiy, A. Stolow, M. S. Schuurman, *J. Am. Chem. Soc.* **2011**, *133*, 16451.

[181] H. Hippler, M. Olzmann, O. Schalk, A.-N. Unterreiner, *Z. Phys. Chem.* **2005**, *219*, 389.

[182] J. D. Idol, C. W. Roberts, E. T. McBee, *J. Org. Chem.* **1955**, *20*, 1743.

[183] C. Sandorfy, *Int. J. Quantum Chem.* **1981**, *19*, 1147.

[184] M. Schreiber, M. R. Silva-Junior, S. P. A. Sauer, W. Thiel, *J. Chem. Phys.* **2008**, *128*, 134110.

[185] T. Baer, W. L. Hase, *Unimolecular reaction dynamics: theory and experiments*, The international series of monographs on chemistry ; 31, Oxford Univ. Pr., New York [u.a.] **1996**.

[186] S. DePaul, D. Pullman, B. Friedrich, *J. Phys. Chem.* **1993**, *97*, 2167.

[187] T. J. A. Wolf, *Untersuchungen zur Relaxationsdynamik des zweiten elektronisch angeregten Zustands halogenierter Cyclopentadiene*, Universität Karlsruhe (TH) **2009**, diploma thesis.

[188] P. Derrick, L. Åsbrink, O. Edqvist, B.-Ö. Jonsson, E. Lindholm, *Int. J. Mass. Spectrom. Ion. Phys.* **1971**, *6*, 203.

[189] B. E. Applegate, T. A. Miller, T. A. Bearckholtz, *J. Chem. Phys* **2001**, *114*, 4855.

[190] B. E. Applegate, A. J. Bezant, T. A. Miller, *J. Chem. Phys* **2001**, *114*, 4869.

[191] S. Zilberg, Y. Haas, *J. Am. Chem. Soc.* **2002**, *124*, 10683.

[192] A. E. Boguslavskiy, M. S. Schuurman, D. Townsend, A. Stolow, *Faraday Discussions* **2011**, *150*, 419.

[193] I. Benjamin, *J. Chem. Phys.* **1995**, *103*, 2459.

[194] I. Chorny, J. Vieceli, I. Benjamin, *J. Chem. Phys.* **2002**, *116*, 8930.

[195] J. Ammer, C. F. Sailer, E. Riedle, H. Mayr, *J. Am. Chem. Soc.* **2012**, *134*, 11481.

[196] C. F. Sailer, B. P. Fingerhut, S. Thallmair, C. Nolte, J. Ammer, H. Mayr, R. de Vivie-Riedle, I. Pugliesi, E. Riedle, *J. Am. Chem. Soc.* **2012**, *in revision*.

[197] M. E. Casida, F. Gutierrez, J. Guan, F.-X. Gadea, D. Salahub, J.-P. Daudey, *J. Chem. Phys.* **2000**, *113*, 7062.

[198] A. Dreuw, J. L. Weisman, M. Head-Gordon, *J. Chem. Phys.* **2003**, *119*, 2943.

[199] K. Sadeghian, M. Schütz, *J. Am. Chem. Soc.* **2007**, *129*, 4068.

[200] J. P. Perdew, A. Zunger, *Phys. Rev. B* **1981**, *23*, 5048.

[201] D. Rappoport, F. Furche, *J. Am. Chem. Soc.* **2004**, *126*, 1277.

[202] H. J. V. Tyrrell, K. R. Harris, *Diffusion in liquids*, Butterworths monographs in chemistry, Butterworths, London **1984**.

[203] O. Schalk, A. N. Unterreiner, *Phys. Chem. Chem. Phys.* **2010**, *12*, 655.

[204] O. Schalk, A.-N. Unterreiner, *Z. Phys. Chem.* **2011**, *225*, 927.

[205] O. Schalk, P. Hockett, *Chemical Physics Letters* **2011**, *517*, 237.

[206] T. J. A. Wolf, D. Voll, C. Barner-Kowollik, A.-N. Unterreiner, *Macromolecules* **2012**, *45*, 2257.

[207] M. Kaur, A. K. Srivastava, *J. Macromol. Sci., Part C: Polym. Rev.* **2002**, *42*, 481.

[208] N. S. Allen, in *Photochemistry*, vol. 36, The Royal Society of Chemistry **2007**, 232.

[209] M. A. Tasdelen, Y. Yagci, *Aust. J. Chem.* **2011**, *64*, 982.

[210] M. Thiel, M. Hermatschweiler, *Optik & Photonik* **2011**, *6*, 36.

[211] A. Biswas, I. S. Bayer, A. S. Biris, T. Wang, E. Dervishi, F. Faupel, *Adv. Colloid Interface Sci.* **2012**, *170*, 2.

[212] F. A. Rueggeberg, *Dent. Mater.* **2011**, *27*, 39.

[213] J. P. Fisher, D. Dean, P. S. Engel, A. G. Mikos, *Annu. Rev. Mater. Res.* **2001**, *31*, 171.

[214] K. S. Anseth, A. T. Metters, S. J. Bryant, P. J. Martens, J. H. Elisseeff, C. N. Bowman, *J. Controlled Release* **2002**, *78*, 199.

[215] J. Ge, M. Trujillo, J. Stansbury, *Dent. Mater.* **2005**, *21*, 1163.

[216] K. Anseth, S. Newman, C. Bowman, in *Biopolymers II, Advances in Polymer Science*, vol. 122, N. Peppas, R. Langer, eds., Springer Berlin / Heidelberg **1995**, 177.

[217] T. Corrales, F. Catalina, C. Peinado, N. Allen, *J. Photochem. Photobiol., A* **2003**, *159*, 103.

[218] F. D. Lewis, R. T. Lauterbach, H. G. Heine, W. Hartmann, H. Rudolph, *J. Am. Chem. Soc.* **1975**, *97*, 1519.

[219] G. Amirzadeh, W. Schnabel, *Makromol. Chem.* **1981**, *182*, 2821.

[220] E. Andrzejewska, D. Zych-Tomkowiak, M. Andrzejewski, G. L. Hug, B. Marciniak, *Macromolecules* **2006**, *39*, 3777.

[221] D. P. Specht, P. A. Martic, S. Farid, *Tetrahedron* **1982**, *38*, 1203.

[222] X. Allonas, J. Fouassier, M. Kaji, M. Miyasaka, T. Hidaka, *Polymer* **2001**, *42*, 7627.

[223] D. K. Balta, N. Cetiner, G. Temel, Z. Turgut, N. Arsu, *J. Photochem. Photobiol., A* **2008**, *199*, 316.

[224] M. A. El Sayed, *J. Chem. Phys.* **1963**, *38*, 2834.

[225] Z. Pawelka, E. S. Kryachko, T. Zeegers-Huyskens, *Chem. Phys.* **2003**, *287*, 143.

[226] T. J. A. Wolf, J. Fischer, M. Wegener, A.-N. Unterreiner, *Opt. Lett.* **2011**, *36*, 3188.

[227] G. Yilmaz, A. Tuzun, Y. Yagci, *J. Polym. Sci. A Polym. Chem.* **2010**, *48*, 5120.

[228] A. Jankowiak, P. Kaszynski, *J. Org. Chem.* **2009**, *74*, 7441.

[229] M. Lipson, N. J. Turro, *J. Photochem. Photobiol., A* **1996**, *99*, 93.

[230] N. K. Shrestha, E. J. Yagi, Y. Takatori, A. Kawai, Y. Kajii, K. Shibuya, K. Obi, *J. Photochem. Photobiol., A* **1998**, *116*, 179.

[231] B. Seidl, R. Liska, G. Grabner, *J. Photochem. Photobiol., A* **2006**, *180*, 109.

[232] S. Jockusch, I. V. Koptyug, P. F. McGarry, G. W. Sluggett, N. J. Turro, D. M. Watkins, *J. Am. Chem. Soc.* **1997**, *119*, 11495.

[233] S. Jockusch, M. S. Landis, B. Freiermuth, N. J. Turro, *Macromolecules* **2001**, *34*, 1619.

[234] N. J. Turro, V. Ramamurthy, J. C. Scaiano, *Modern Molecular Photochemistry of Organic Molecules*, University Science Books, Sausalito **2010**.

[235] R. Morales-Cueto, M. Esquivelzeta-Rabell, J. Saucedo-Zugazagoitia, J. Peon, *J. Phys. Chem. A* **2007**, *111*, 552.

[236] E. Collado-Fregoso, J. S. Zugazagoitia, E. F. Plaza-Medina, J. Peon, *J. Phys. Chem. A* **2009**, *113*, 13498.

[237] E. F. Plaza-Medina, W. Rodríguez-Córdoba, R. Morales-Cueto, J. Peon, *J. Phys. Chem. A* **2011**, *115*, 577.

[238] C. Ma, Y. Du, W. M. Kwok, D. L. Phillips, *Chem.-Eur. J.* **2007**, *13*, 2290.

[239] E. E. Fileti, S. Canuto, *Int. J. Quantum Chem.* **2005**, *104*, 808.

[240] A. Klamt, G. Schüürmann, *J. Chem. Soc., Perkin Trans. 2* **1993**, 799.

[241] S. W. Hell, J. Wichmann, *Opt. Lett.* **1994**, *19*, 780.

[242] T. A. Klar, S. Jakobs, M. Dyba, A. Egner, S. W. Hell, *Proc. Natl. Acad. Sci. U. S. A.* **2000**, *97*, 8206.

[243] M. Dyba, S. W. Hell, *Phys. Rev. Lett.* **2002**, *88*, 163901.

[244] S. W. Hell, *Science* **2007**, *316*, 1153.

[245] S. W. Hell, *Nat. Methods* **2009**, *6*, 24.

[246] J. Fischer, M. Wegener, *Opt. Mater. Express* **2011**, *1*, 614.

[247] S. R. Trenor, A. R. Shultz, B. J. Love, T. E. Long, *Chem. Rev.* **2004**, *104*, 3059.

[248] C. K. Nguyen, C. E. Hoyle, T. Y. Lee, S. Jönsson, *Eur. Pol. J.* **2007**, *43*, 172.

[249] G. Yilmaz, B. Aydogan, G. Temel, N. Arsu, N. Moszner, Y. Yagci, *Macromolecules* **2010**, *43*, 4520.

[250] J. Lalevée, N. Blanchard, M. A. Tehfe, C. Fries, F. Morlet-Savary, D. Gigmes, J. P. Fouassier, *Polym. Chem.* **2011**, *2*, 1077.

[251] L. Li, R. R. Gattass, E. Gershgoren, H. Hwang, J. T. Fourkas, *Science* **2009**, *324*, 910.

[252] T. F. Scott, B. A. Kowalski, A. C. Sullivan, C. N. Bowman, R. R. McLeod, *Science* **2009**, *324*, 913.

[253] T.-S. Kim, J. Kim, S. Bae, Y.-Y. Choi, S. Kim, *Ind. Eng. Chem. Res.* **2007**, *46*, 4799.

[254] U. Brackmann, *Lambdachrome®Laser Dyes*, 1 ed., Lambda Physik GmbH **1986**.

[255] J. Fischer, T. Ergin, M. Wegener, *Opt. Lett.* **2011**, *36*, 2059.

[256] F. G. Bordwell, G. E. Drucker, H. E. Fried, *J. Org. Chem* **1981**, *46*, 632.

[257] G. E. Hawkes, R. A. Smith, J. D. Roberts, *J. Org. Chem.* **1974**, *39*, 1276.

List of publications

Publications

- D. T. Thielemann, M. Klinger, T. J. A. Wolf, Y. Lan, W. Wernsdorfer, M. Busse, P. W. Roesky, A.-N. Unterreiner, A. K. Powell, P. C. Junk, G. B. Deacon, Novel Lanthanide-Based Polymeric Chains and Corresponding Ultrafast Dynamics in Solution, *Inorg. Chem.* **2011**, *50*, 11990.

- T. J. A. Wolf, J. Fischer, M. Wegener, A.-N. Unterreiner, Pump-probe spectroscopy on photoinitiators for stimulated-emission-depletion optical lithography, *Opt. Lett.* **2011**, *36*, 3188.

- T. J. A. Wolf, D. Voll, C. Barner-Kowollik, A.-N. Unterreiner, Elucidating the Early Steps in Photoinitiated Radical Polymerization via Femtosecond Pump-Probe Experiments and DFT Calculations, *Macromolecules* **2012**, *45*, 2257.

- P. D. Dau, H.-T. Liu, J.-P. Yang, M.-O. Winghart, T. J. A. Wolf, A.-N. Unterreiner, P. Weis, Y.-R. Miao, C.-G. Ning, M. M. Kappes, L.-S. Wang, Resonant tunneling through the repulsive Coulomb barrier of a quadruply charged molecular anion, *Phys. Rev. A* **2012**, *85*, 064503.

Conference contributions

- T. J. A. Wolf, A.-N. Unterreiner, The ultrafast photochemistry of fully halogenated cyclopentadienes in solution, *Bunsentagung 2011, 110th Annual German Conference on Physical Chemistry*, **2011**, Talk.

- T. J. A. Wolf, J. Fischer, D. Voll, M. Wegener, C. Barner-Kowollik, A.-N. Unterreiner, The interplay of different relaxation channels in the excited state dynamics of photoinitiators, *XVIII International Conference on Ultrafast Phenomena*, **2012**, Poster.

Curriculum Vitae

Personal data

Name:	Thomas Jakob Arcangelo Wolf
Date of birth:	September 28, 1983
Place of birth:	Filderstadt
Nationality:	German
Family status	Unmarried

School education

1990 - 1994	Gerhardinger Schule Regensburg
1994 - 1996	Albertus Magnus Gymnasium Regensburg
1996 - 2003	Uhlandgymnasium Tübingen
2003	Abitur

Civilian service

2003 - 2004	EMS, German Red Cross, Tübingen

Academic studies

2004 - 2009	Studies in chemistry, University of Karlsruhe (TH)
2006	Intermediate diploma in chemistry
05/2009	Diploma in chemistry, University of Karlsruhe (TH)
Since 07/2009	Research assistant at the Institute of Molecular Physical Chemistry, Karlsruhe Institute of Technology (KIT)
03/2010 - 02/2012	Scholarship by the Fonds der Chemischen Industrie (FCI)

Danksagung

Ich möchte mich im folgenden bei allen bedanken, die mir das Verfassen der vorliegenden Arbeit ermöglicht oder zu ihrem Gelingen beigetragen haben:

Ich möchte mich bei Herrn PD Dr. A.-N. Unterreiner für die Aufgabenstellung, Unterstützung in jeglicher Hinsicht sowie für hilfreiche Diskussionen herzlich bedanken.

Mein besonderer Dank gilt Herrn Dr. O. Schalk für viele erhellende Diskussionen, Anregungen, Ideen und nicht zuletzt für die Ermöglichung meines Messaufenthalts bei der Stolow-Gruppe in Ottawa.

I thank Prof. Dr. A. Stolow for the kind invitation to visit his group in Ottawa.

Ich möchte mich bei Herrn Dr. J. Fischer, Herrn Dr. D. Voll, Herrn Prof. Dr. M. Wegener und Herrn Prof. Dr. C. Barner-Kowollik für angenehme und erfolgreiche Kooperationsprojekte bedanken.

Ein herzlicher Dank geht auch an den Fonds der Chemischen Industrie für die finanzielle Unterstützung.

Vielen Dank an Frau Dr. M. Klinger, Frau H. Ernst und Herrn Y. Liang für vielerlei Hilfestellungen und die gute Atmosphäre im Femto-Arbeitskreis.

Herzlichen Dank an Herrn D. Kelly für die Durchsicht meiner Dissertation und Hilfe in sprachlichen Fragen.

Dank geht auch an Herrn Dr. T. Bentz, Frau Dr. A. Busch, Herrn Dr. S. Dürrstein, Herrn P. Friese, Frau L. Genthner, Herrn J. Hetzler, Frau P. Hibomvschi, Frau C. Hüllemann, Frau Dr. C. Kappler, Herrn J. Kiecherer, Herrn Prof. Dr. M. Olzmann, Herrn M. Pfeifle, Herrn Dr. J. Sommerer, Herrn Dr. M. Szöri und Herrn Dr. O. Welz für eine angenehme und produktive Arbeitsatmosphäre, Unterstützung in vielerlei Hinsicht und genug Kuchen.

Außerdem möchte ich mich bei Claudia, Hans und Johannes bedanken, ohne die ich mein Chemiestudium niemals geschafft hätte.

Zuletzt gilt mein Dank meinen Eltern, ohne deren nicht nur finanzielle Unterstützung diese Arbeit nicht zustande gekommen wäre.